Green Sustainable Energy

As the world struggles with sustainable practices and policies regarding environmental protection, green sustainable energy is a key player. The chemistries involving energy production must be efficient but also must evolve and change to meet new restrictions and footprint guidelines. Rather than only being seen as a necessary evil, energy through green sustainable energy (GSE) must become a key piece in the sustainability puzzle. The green sustainable energies presented in this book will demonstrate that progress in science can and should be leading contributors in discussions on environmental science and chemistry.

This book:

- Explains the necessary role of energy in the sustainability of the world in the 21st century
- Recognizes past practices and future potential, guided by global demand and the four drivers: economic, scientific, regulatory, and environmental
- Presents a much-needed multi-dimensional approach to the subject
- Demonstrates that green sustainable energies can and should be leading contributors in discussions on environmental science and chemistry
- Highlights new products, processes, applications, and developments in green sustainable energy, which demonstrates how sustainability is adapting in the new age

Sustainability: Contributions through Science and Technology
Series Editors:
Thomas P. Umile, Ph.D, Villanova University, Pennsylvania, USA
William M. Nelson, US Army ERDC, USA

Preface to the Series

Sustainability is rapidly moving from the wings to center stage. Overconsumption of non-renewable and renewable resources, as well as the concomitant production of waste has brought the world to a crossroads. Green chemistry, along with other green sciences technologies, must play a leading role in bringing about a sustainable society. The Sustainability: Contributions through Science and Technology series focuses on the role science can play in developing technologies that lessen our environmental impact. This highly interdisciplinary series discusses significant and timely topics ranging from energy research to the implementation of sustainable technologies. Our intention is for scientists from a variety of disciplines to provide contributions that recognize how the development of green technologies affects the triple bottom line (society, economic, and environment). The series will be of interest to academics, researchers, professionals, business leaders, policy makers, and students, as well as individuals who want to know the basics of the science and technology of sustainability.

Michael C. Cann

Green Chemistry for Environmental Sustainability *Edited by Sanjay Kumar Sharma, Ackmez Mudhoo, 2010*

Microwave Heating as a Tool for Sustainable Chemistry *Edited by Nicholas E. Leadbeater, 2010*

Green Organic Chemistry in Lecture and Laboratory *Edited by Andrew P. Dicks, 2011*

A Novel Green Treatment for Textiles: Plasma Treatment as a Sustainable Technology *C. W. Kan, 2014*

Environmentally Friendly Syntheses Using Ionic Liquids *Edited by Jairton Dupont, Toshiyuki Itoh, Pedro Lozano, Sanjay V. Malhotra, 2015*

Catalysis for Sustainability: Goals, Challenges, and Impacts *Edited by Thomas P. Umile, 2015*

Nanocellulose and Sustainability: Production, Properties, Applications, and Case Studies *Edited by Koon-Yang Lee, 2017*

Sustainability of Biomass through Bio-based Chemistry *Edited by Valentin Popa, 2021*

Nanotechnologies in Green Chemistry and Environmental Sustainability, 2022

Towards Sustainability in the Wine Industry by Valorization of Waste Products: Bioactive Extracts *Edited by Patricia Joyce Pamela Zorro Mateus and Siby Inés Garcés Polo, 2023*

Climate Change and Carbon Recycling: Surface Chemistry Applications, *K.S. Birdi*

Green Sustainable Energy

William M. Nelson

CRC Press
Taylor & Francis Group
Boca Raton London New York

CRC Press is an imprint of the
Taylor & Francis Group, an **informa** business

First edition published 2026
by CRC Press
2385 NW Executive Center Drive, Suite 320, Boca Raton FL 33431

and by CRC Press
4 Park Square, Milton Park, Abingdon, Oxon, OX14 4RN

CRC Press is an imprint of Taylor & Francis Group, LLC

ISBN: 978-1-032-52593-8 (hbk)
ISBN: 978-1-032-52601-0 (pbk)
ISBN: 978-1-003-40744-7 (ebk)

DOI 10.1201/9781003407447

Typeset in Sabon
by Deanta Global Publishing Services, Chennai, India

Contents

About the author

William M. Nelson is an organic chemist who has been involved in green chemistry since 1995. Growing up within a military family, he has been able to see many parts of the world, which heightened his commitment to preserving the natural environment and promoting sustainability through chemistry. He earned his doctorate in organic chemistry from The Johns Hopkins University, during which time he studied the synthesis and photobiology of analogs of the environmental carcinogen benzo[a]pyrene. Dr. Nelson has worked in industry as a research chemist, in government (with both the Illinois EPA and US EPA), in environmental protection and pollution prevention, and in education (teaching and directing research). He was a Research Physical Scientist in the Environmental Chemistry Branch of the Environmental Laboratory at Engineering Research and Development Center (ERDC) for the US Army Corps of Engineers, where his work involves organic synthesis and photochemistry.

Introduction

A universe of energy

> ....*neither energy nor mass can be destroyed; rather, both remain constant* [1]

One theory about our universe is that it began with an explosion – the Big Bang. Starting from extremely high density and temperature, space expanded, the universe cooled, and the simplest elements formed. Gravity gradually drew matter together to form the first stars and the first galaxies [2]. Within this matrix was stored the mass and energy that we draw upon today (see Figure 0.1). The challenge has been and will be how to sustainably extract and utilize this energy. A planetary-scale Anthropocene [3–6] time interval is the current era we are in, having begun ~1950 CE [7]. Eras, or epochs, reflect changes in human endeavors resulting in environmental change, and the current shows an increasingly significant and overwhelming global impact. The importance of energy (for our society, for

Figure 0.1 Creation of our world within the universe of energy.

DOI: 10.1201/9781003407447-1

1

our civilization, and for our world) requires that we identify the sources and effects of the energy we use and select those most conducive to our world's survival.

Since 1950 CE, energy supplies have come from a diversified portfolio of energy sources (coal, oil, gas, nuclear, and renewables). Fossil fuels power more than 80% of the global economy [7]. In total, 60% of all human-produced energy has been consumed since 1950 CE. Simultaneously, the Earth's oceans have stored solar energy trapped through the increases in anthropogenic greenhouse gases [8].

Overlay this information with the fact that the human population has exceeded historical natural limits, with (1) the development of new energy sources, (2) technological developments aiding productivity, education, and health, and (3) an unchallenged position on top of food webs. The human species remains Earth's only species to employ technology to change the sources, uses, and distribution of energy forms, including the release of geologically trapped energy (i.e., coal, petroleum, uranium) [7].

0.1 UNIVERSAL ENERGY

If we consider the amount of energy packed in the nucleus of a single uranium atom, the energy that has been continuously radiating from stars for billions of years, or the fact that there are 10^{80} particles in the observable universe, it seems that the total energy in the universe must be an immense quantity [9]. The existence of so much energy seems like a daunting task to identify, but the challenge remains to utilize the types/amounts of energy needed to maintain civilization.

Energy is the substance that animates all things in the universe. In physics, energy is the quantitative property that is transferred to a body or to a physical system, and identified by performance of work and in the form of heat and light. The total energy of any system can be subdivided into potential energy, kinetic energy, or combinations of the two in various forms. While these two categories are sufficient to describe all forms of energy, it is often convenient to combine potential and kinetic energy when referring to the total energy of the system. There also exist processes in the universe that transform energy from one form into another: e.g., mechanical processes, thermal processes, electrical processes, nuclear processes, etc. However, and this may seem unbelievable, all of these processes leave the total amount of energy in the universe invariant. In other words, whenever, and however, energy is transformed from one form into another, it is always conserved [10]. Energy is also equivalent to mass, and this mass can be calculated. Reciprocally, mass is also equivalent to a certain amount of energy, as described in mass–energy equivalence. The formula $E = mc^2$, derived by Albert Einstein (1905), quantifies the relationship between relativistic mass and energy within the concept of special relativity [11].

Sustainability of energy is how these concepts can be woven together. It has been shown that sustainability is a universal quantum-statistical phenomenon, which occurs during propagation of electromagnetic waves, in such diverse media as waveguides, metamaterials, or biological tissues [12]. Since energy is a fixed quantity that cannot be created or destroyed, but only transformed, the total amount of energy can be represented as the sum of energy in the universe. If sustainability can exist as a fundamental property of nature, then we must be able to apply it in the realm of energy [13]. This view of energy sustainability is the integrating phenomenon that will guide this book.

0.1.1 Form and transformation

In cosmology and astronomy, the phenomena of stars, nova, supernova, quasars, and gamma-ray bursts are the universe's highest-output energy transformations of matter. All stellar phenomena (including solar activity) are driven by various kinds of energy transformations. Energy in such transformations is either from gravitational collapse of matter (usually molecular hydrogen) into various classes of astronomical objects (stars, black holes, etc.), or from nuclear fusion (of lighter elements, primarily hydrogen). The nuclear fusion of hydrogen in the Sun also releases another store of potential energy which was created at the time of the Big Bang. This means that hydrogen represents a store of potential energy that can be released by fusion. Such a fusion process comes from gravitational collapse of hydrogen clouds when they produce stars, and some of the fusion energy is transformed into sunlight.

Energy transformations in the universe have been occurring since the Big Bang, whenever a triggering mechanism is available. Familiar examples of such processes include nucleosynthesis, a process ultimately using the gravitational potential energy released from the gravitational collapse of supernovae to "store" energy in the creation of heavy isotopes (such as uranium and thorium), and nuclear decay, a process in which energy is released that was originally stored in these heavy elements, before they were incorporated into the Solar System and the Earth [14].

0.1.2 Total amount of energy in the universe

The zero-energy universe hypothesis proposes that the total amount of energy in the universe is exactly zero: its amount of positive energy in the form of matter is exactly canceled out by its negative energy in the form of gravity. Some physicists, such as Lawrence Krauss, Stephen Hawking, or Alexander Vilenkin, call or called this state "a universe from nothingness," although the zero-energy universe model requires both a matter field with positive energy and a gravitational field with negative energy to exist [15]. In his book *Brief Answers to the Big Questions*, Hawking explains:

The laws of physics demand the existence of something called "negative energy." This is the principle behind what happened at the beginning of the universe. When the Big Bang produced a massive amount of positive energy, it simultaneously produced the same amount of negative energy. In this way, the positive and the negative add up to zero, always. It's another law of nature [16].

Both the fundamental constants that describe the laws of physics and the cosmological parameters that determine the properties of our universe must fall within a range of values for the cosmos to maintain itself and ultimately support life. Some parameters are allowed to vary. The total energy density of the universe includes the contribution from vacuum energy, the dark matter contribution, and the amplitude of primordial density fluctuations. On smaller scales, stars and planets must be able to form and function. The stars must be sufficiently long-lived, have high enough surface temperatures, and have smaller masses than their host galaxies. The planets must be massive enough to hold onto an atmosphere, yet small enough to remain non-degenerate, and contain enough particles to support a biosphere of sufficient complexity. These processes will require a gravitational structure constant, the fine structure constant, and composite parameters that allow nuclear reaction rates. During this process will appear the production of carbon and life as we know it. Finally, the universe can involve new types of astrophysical processes that can generate energy and thereby support habitability [17].

0.1.3 Capturing energy by the earth

Sunlight is the main input to Earth's energy budget which accounts for its temperature and climate stability. Sunlight may be stored as gravitational potential energy after it strikes the Earth, as when water evaporates from oceans and is deposited upon mountains (where, after being released at a hydroelectric dam, it can be used to drive turbines or generators to produce electricity) [18]. Sunlight also drives most weather phenomena, save a few exceptions, like those generated by volcanic events for example. An example of a solar-mediated weather event is a hurricane, which occurs when large unstable areas of warm ocean, heated over months, suddenly give up some of their thermal energy to power a few days of violent air movement.

In geology, continental drift, mountain ranges, volcanoes, and earthquakes are phenomena that can be explained in terms of energy transformations in the Earth's interior ("Earth's Energy Budget") [19] (see Figure 0.2). Meteorological phenomena like wind, rain, hail, snow, lightning, tornadoes, and hurricanes are all a result of energy transformations in our atmosphere brought about by solar energy.

In a slower process, the radioactive decay of atoms in the core of the Earth releases heat. This thermal energy drives plate tectonics and may lift mountains, via orogenesis. This slow lifting represents a kind of gravitational potential energy storage of the thermal energy, which may later be transformed into active kinetic energy during landslides, after a triggering event. Earthquakes also release stored elastic potential energy in rocks, a store that has been produced ultimately from the same radioactive heat sources [20].

0.2 EARTH ENERGY

Earth's energy budget depends on many factors, such as atmospheric aerosols, greenhouse gases, the planet's surface albedo (reflectivity), clouds, vegetation, land use patterns, and more (Figure 0.2). When the incoming and outgoing energy fluxes are in balance, Earth is in radiative equilibrium and the climate system will be relatively stable. Global warming occurs when the Earth receives more energy than it gives back to space, and global cooling takes place when the outgoing energy is greater [22].

Water plays a major role in the storage and transfer of energy in the Earth system. The major role water plays is a result of water's prevalence, high heat capacity, and the fact that phase changes of water occur regularly on Earth [23]. The Sun provides the energy that drives the water cycle on Earth.

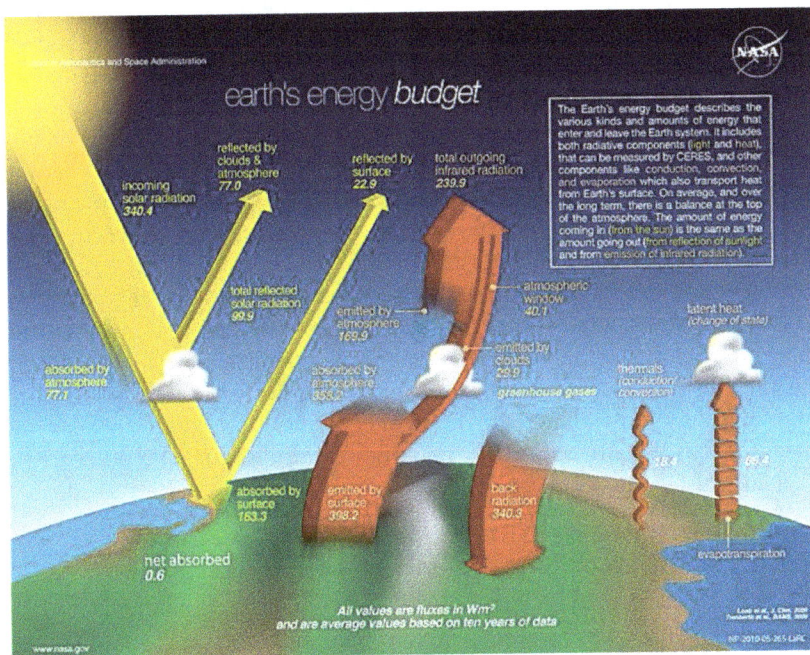

Figure 0.2 Earth's energy budget [21, 22].

Photosynthesis is a significant avenue for energy capture: An estimated 140 TW (or around 0.08%) of incident energy gets captured by photosynthesis, giving energy to plants to produce biomass. A similar flow of thermal energy is released over the course of a year when plants are used as food or fuel [24].

0.2.1 Earth energy flows

Despite the enormous transfers of energy into and from the Earth, the planet maintains a relatively constant temperature because, as a whole, there is little net gain or loss. The Earth emits via atmospheric and terrestrial radiation (shifted to longer electromagnetic wavelengths) to space about the same amount of energy as it receives via solar insolation (all forms of electromagnetic radiation).

0.2.1.1 Solar energy

The total amount of energy received per second at the top of Earth's atmosphere (TOA) is measured in watts and is approximately 340 watts per square meter (W/m²). Since the absorption varies with location as well as with diurnal, seasonal, and annual variations, these numbers are multi-year averages obtained from multiple satellite measurements [25].

Of the ~340 W/m² of solar radiation received by the Earth, an average of ~77 W/m² is reflected back to space by clouds and the atmosphere and ~23 W/m² is reflected by the surface albedo, leaving ~240 W/m² of solar energy input to the Earth's energy budget. This amount is called the absorbed solar radiation (ASR). It implies a value of about 0.3 for the mean net albedo of Earth [26].

Generally, absorbed solar energy is converted to different forms of heat energy. Some of this energy is emitted as outgoing longwave radiation (OLR) directly to space, while the rest is first transported through the climate system as radiant and other forms of thermal energy. For example, indirect emissions occur following heat transport from the planet's surface layers (land and ocean) to the atmosphere via evapotranspiration and latent heat fluxes or conduction/convection processes [27]. Ultimately, all of the outgoing energy is radiated in the form of longwave radiation back into space. OLR is usually defined as outgoing energy leaving the planet, most of which is in the infrared band.

0.2.1.2 Earth's internal heat sources and other small effects

The geothermal heat flow from the Earth's interior is estimated to be 47 terawatts (TW) and split approximately equally between radiogenic heat and heat left over from the Earth's formation. This corresponds to an average flux of 0.087 W/m² and represents only 0.027% of Earth's total energy budget at the surface, being dwarfed by the 173,000 TW of incoming solar radiation [28].

Human production of energy is even lower at an estimated average continuous heat flow of about 18 TW. However, human consumption is growing rapidly and energy production using fossil fuels also produces an increase in atmospheric greenhouse gases, leading to a more than 20 times larger imbalance in the incoming/outgoing flows that originate from solar radiation [29].

0.2.2 Earth energy sources

The energy of the earth is contained in many resources, from which is harvested useful energy. These terrestrial energy sources mainly include coal, geothermal energy, wind energy, biomass, petroleum, and nuclear energy. These sources are transformed into electricity and fuels for human use. The various energy types are shown in Table 0.1 [30].

0.2.3 Energy's role in earth homeostasis

Solar energy is the driver of many Earth System processes. This energy flows into the Atmosphere and heats this system. It also heats the Hydrosphere and the land surface of the Geosphere and fuels many processes in the

Table 0.1 Some forms of energy

Type of energy	Description
Mechanical	The sum of macroscopic translational and rotational kinetic and potential energies
Electric	Energy from influence of electric fields
Magnetic	Energy from influence of magnetic fields
Gravitational	Energy from influence of gravitational fields
Chemical	Energy due to chemical bonds
Ionization	Energy that binds an electron to its atom
Nuclear	Energy that binds nucleons to form the atomic nucleus
Chromodynamic	Energy that binds quarks to form hadrons
Elastic	Potential energy due to the deformation of a material (or its container) having a restorative force to return to its original shape
Mechanical wave	Kinetic and potential energy in an elastic material due to a propagating oscillation of matter
Sound wave	Kinetic and potential energy in a material due to a sound-propagated wave
Radiant	Energy stored in the fields of waves propagated by electromagnetic radiation
Rest	Energy due to an object's rest mass
Thermal	Kinetic energy equivalent of mechanical energy from the microscopic motion of particles

Biosphere. Differences in the amount of energy absorbed in different places set the atmosphere and oceans in motion and help determine their overall temperature and chemical structure. These motions, such as wind patterns and ocean currents, redistribute energy throughout the environment. Eventually, the energy that began as sunshine (short-wave radiation) leaves the planet as light reflected by the atmosphere and surface back into space [31].

The Earth maintains a stable average temperature and climate by achieving an earth-atmosphere energy balance, where the energy received from the Sun balances the energy lost by the Earth back into space. Earth's temperature doesn't infinitely rise because the surface and the atmosphere are simultaneously radiating heat to space, which is Earth's energy budget. However, the incoming energy to the Earth and the outgoing energy from the Earth do not actually balance, partially due to the incoming energy from the Sun, which varies with the seasons and changes in the composition of the Earth's atmosphere [32].

The current Earth's energy imbalance (EEI) is being increased due to human activity driving global warming. The absolute value of EEI provides the most fundamental metric underlying global climate change and is more useful than using global surface temperature. EEI can best be estimated from changes in ocean heat content, complemented by radiation measurements from space. Regular satellite and earth measurements are crucial to refining future estimates of EEI. Combining multiple measurements in an optimal way holds considerable promise for estimating EEI and thus assessing the status of global climate change, improving climate syntheses and models, and testing the effectiveness of mitigation actions [33].

0.3 THE WORLD OF ENERGY REQUIREMENTS

Energy has been essential to the development of human civilization for millennia. A valuable metric of energy consumption and utilization is the Kardashev Scale (KS), which was proposed to quantify the relationship between energy consumption and the development of civilizations [34]. Humanity presently stands at Type 0.7276 on the KS. Machine learning models predict energy consumption on a global scale and the position of humanity on the Kardashev Scale through 2060. The result suggests that global energy consumption is expected to reach ~ 887 EJ in 2060, and humanity will become a Type 0.7449 civilization. If energy strategies and technologies remain in the present course, it may take human civilization millennia to become a Type 1 civilization [34].

0.3.1 Uses of energy

Modern societies divide energy use among four different economic sectors: residential, commercial, transportation, and industrial. Heating and cooling homes, lighting office buildings, driving cars and moving freight, and manufacturing products for daily use are all functions that require energy. The demand for energy will increase. In the United States alone, energy consumption is expected to rise 7.3% over the next two decades. Global consumption is expected to increase by 40% over the same time period [35].

0.3.1.1 Home and work

Energy use in homes and commercial buildings is very similar. Rooms are kept at a comfortable temperature, illuminated, water heated for bathing and laundry, and computers, copiers, appliances, and other technologies are kept working [36]. Buildings consume 75% of the electricity and 40% of the total energy used in the United States, accounting for 36% of all U.S. carbon dioxide emissions. National-level analysis showed a 16.7% increase in consumption, likely due to the increased access to natural gas in many states and the increased use of natural gas for major end-use consumption categories in homes (e.g., heating, cooling, water heating) [37].

The amount of energy consumed by specific uses has changed during the 21st century, often dramatically. For decades, more than half of all residential energy use went to space heating and cooling. But U.S. Energy Information Administration (EIA) data show that energy for appliances, electronics, and lighting rose from 24% to 35%, owing to the proliferation of appliances, as well as trends toward larger televisions and other devices [38].

As the global population and living standards rise, it pushes the demand for basic amenities like food, health, and energy resources. Additionally, manufacturing automation has led to mass production and consumption, triggering waste production. The existing linear economy approach has led to increasing waste production and resource depletion, posing significant environmental and public health threats. To overcome these impediments, an alternative model called the circular economy concept has gained popularity in the global industry community [39].

Numerous studies on the nexus between energy consumption in residential and industrial sectors in the United States have shown obvious trends. The U.S. manufacturing and residential consumption are skewed by the industrial and transportation sectors. These sectors have increasing energy consumption, while others are negative. On average, there is a positive growth of energy consumption in the United States depending upon increased manufacturing performance. This trend can be extrapolated to world production [40]. The general trend is shown in Figure 0.3.

0.3.1.2 Transportation

The United States used 27% of total U.S. energy consumption in 2022 for transporting people and goods from one place to another [42]. The

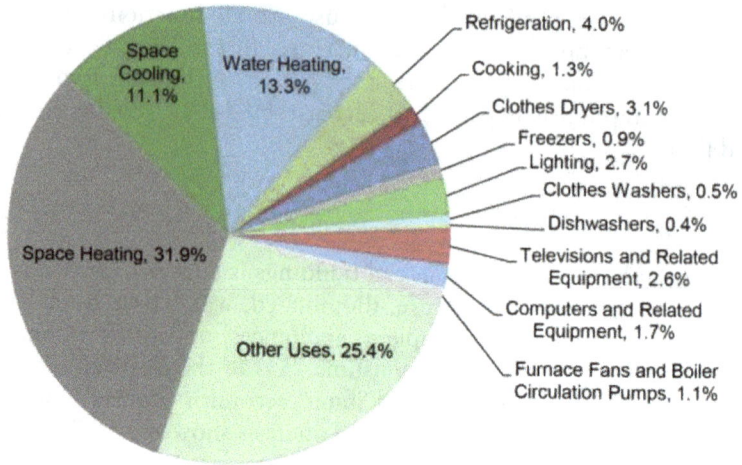

Figure 0.3 U.S. residential energy consumption by end use, 2021 [41].

transportation sector includes many modes, from personal vehicles and large trucks to public transportation (buses, trains) to airplanes, freight trains, ships and barges, and pipelines. By far the largest share is consumed by cars, light trucks, and motorcycles – about 58%, followed by other trucks (23%), aircraft (8%), boats and ships (4%), and trains and buses (3%). Pipelines account for 4%.

During the past century, dependence on vehicles burning petroleum-based fuels has become a defining component of American life. The United States, with less than 5% of the world's population, is home to more than one-fifth of the world's automobiles. During the next 25 years, the total number of miles driven by Americans is projected to grow by about 23%, increasing the demand for fuel. However, advances in efficiency and changes in the types of vehicles purchased will change this pattern. Overall, it appears that gains in vehicle efficiency will be offset by increased miles driven.

In 2022, petroleum products accounted for about 90% of total U.S. transportation sector energy use. Biofuels contributed about 6%, most of which were blended with petroleum fuels (gasoline, diesel fuel, and jet fuel). Natural gas accounted for about 5%, and nearly all was used as a fuel for natural gas pipeline compressors. Electricity use by mass transit systems was less than 1% of total energy consumption by the transportation sector (see Figure 0.4).

Alternative geological sources to conventional petroleum deposits, such as oil from low-permeability geological formations (tight oil), are adding appreciably to the nation's supply of transportation fuels. But liquids from those sources cannot solve the environmental issues associated with burning fossil fuels.

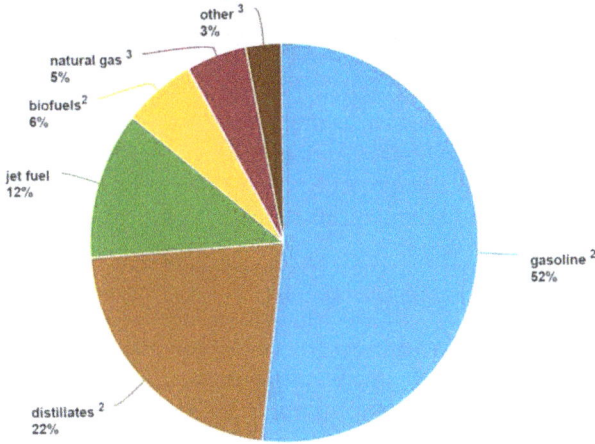

1. Based on energy content
2. Gasoline is motor gasoline
3. Residual fuel oil, lubricants, etc.

Figure 0.4 U.S. transportation energy sources 2022[1] [42].

There are alternatives to oil. Biofuels, which provided about 4.9% of the nation's transportation fuel in 2015, chiefly in the form of ethanol added to gasoline, are an option, but with limits. Corn ethanol production is energy-intensive, reduces the acreage devoted to growing food (in 2015, about 38% of the harvested corn in the United States went to making ethanol and its associated coproducts), and requires substantial amounts of water. Alternative types of vehicles – hybrids, all-electric vehicles, and vehicles powered by hydrogen or natural gas, for example – can contribute to the goal of reducing dependence on oil. However, arriving at an optimal mix of vehicle types may be quite complicated, especially in finding a cost-effective solution. Vehicles designed for different fuels employ different engines, associated technologies, and maintenance procedures. Each requires its own distinctive fuel delivery system. Hydrogen filling stations, for instance, entail novel technologies and would require widespread deployment before hydrogen-powered cars became a practical alternative for most drivers. Potential mainstream adopters of plug-in vehicles require additional encouragement, information, and incentives to overcome perceived barriers to ownership and use.

The best strategy for reducing demand for petroleum may be encouraging the trend toward improved efficiency of conventional vehicles. Much of that improvement has resulted from the Corporate Average Fuel Efficiency (CAFE) standards, initially adopted in 1975 and revised in 2007. The standards have recently been made more stringent, with fuel economy requirements increasing annually between 2012 and 2025. A study of the 2017–2025 CAFE standards found that a mid-size vehicle with advanced conventional gasoline engine technologies is likely to meet the increased standards. Regulations aimed at implementing the first standards for model years 2014 to 2018 medium- and heavy-duty trucks – based on analyses from the National Academies of Sciences, Engineering, and Medicine – are currently in effect, and regulations for beyond 2018 have been released [43].

0.3.1.3 Industry

The industrial sector accounted for 37% (166 EJ) of global energy use in 2022, compared to 34% in 2002. Growth in energy consumption over the past decade has been driven largely by continued rising production in energy-intensive industry subsectors [44]. That's understandable in view of the wide range of activity in this economic sector. Every product on which we rely – from gasoline and automobiles to food, buildings, machinery, and appliances – takes energy to produce. The use of energy in industry affects every single citizen directly through the cost of goods and services, the quality of manufactured products, the strength of the economy, and the availability of jobs.

Industrial energy consumption is still dominated by fossil fuels, in particular coal, and accounts for about a quarter of energy-related CO_2 emissions. As the global economy and population grow, so will the demand for materials and goods, increasing the importance of understanding which technologies and strategies can support the sustainable manufacture, use, and disposal of indispensable commodities.

A few industries use a very large share of energy in the industrial sector. Petroleum refining is the principal consumer, with the chemical industry a close second. Those users, plus the paper and metal industries, account for 78% of total industrial energy use [44].

Modest improvements have already been made in energy efficiency and in renewable energy uptake, and some positive steps have been taken in the areas of international collaboration and innovation. However, progress is occurring far too slowly. Greater material and energy efficiency, more rapid uptake of low carbon fuels, and faster development and deployment of near zero-emission (NZE) production processes – including carbon capture, utilization and storage (CCUS), and hydrogen – are all needed if meaningful progress toward NZE Scenario milestones is to be made by 2030 [44].

0.3.2 Energy production

Energy production refers to how much primary energy a country extracts from nature. This is the total of all of the harvested primary fuels and primary energy flows. Production ignores both imports and exports and

Country's Energy Flows

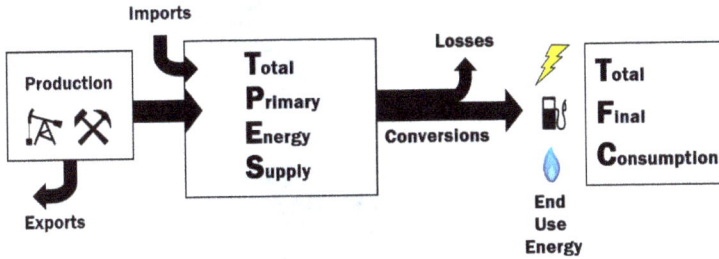

Figure 0.5 How production becomes total primary energy supply becomes total final consumption [45, 46].

sums up what's extracted from nature. The primary energy available for use by a country after imports and exports is the total primary energy supply (TPES) and can be thought of as an energy mix. See Figure 0.5 for a flow-chart that explains how production becomes TPES and eventually the total final consumption (TFC).

Energy production would include:

- Any coal, oil, or natural gas that is extracted from the ground in that country (but not the fossil fuels that are imported).
- Any hydropower, wind, geothermal, tidal, or solar power extracted from nature.
- Electricity from nuclear power plants, rather than the energy held in uranium that's been mined because there's so much more energy in the uranium than any of the other sources.

In 2022, coal accounted for roughly 35.8% of the global power mix, while natural gas followed with a 22% share. China, India, and the United States accounted for the largest share of coal used for electricity generation in 2021 [47] (see Figure 0.6).

0.4 ENERGY STEWARDSHIP

Energy transitioning toward sustainability involves many working parts: researchers, businesses, and policymakers. They all explore and usefully improve energy systems and energy consumption behavior, both to reflect the reality of climate change and related environmental degradation and to adapt to the expanding periphery of renewable energy technologies. There are potential policy pathways to the necessary transformation in societal energy consumption, usage, and behavior. Solutions must include energy

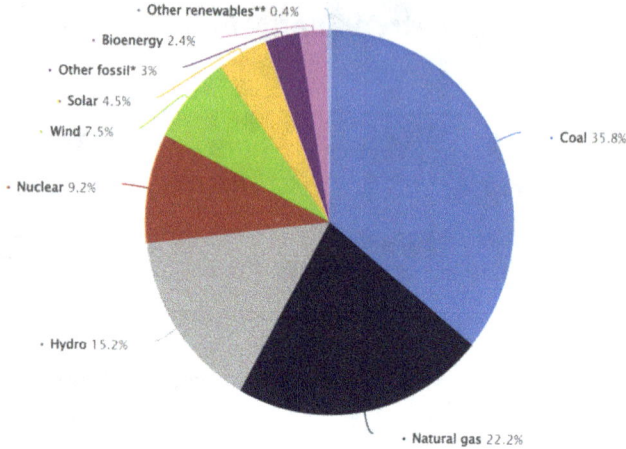

Figure 0.6 World energy production [47].

efficiency, energy security, the role of political leadership, green public policy, and the transition to renewable energy sources. Energy stewardship should be the driver of this transformation [48].

0.4.1 Technology

Energy stewardship involves safe and reliable production, transmission, and storage of energy. It should be used in manufacturing, transportation, communications, heating, and cooling for domestic and commercial purposes. All involved should work to improve energy efficiency and conservation by educating and raising awareness about opportunities for improvement. The Energy Stewardship philosophy is centered around sharing learning among those involved.

A primary role for science and technology must be identification of energy sources within the universe of energy. Most of the energy in the Earth's system, as we showed earlier, comes from just a few sources: solar energy, gravity, radioactive decay, and the rotation of the Earth. Solar energy drives many surface processes such as winds, currents, the hydrologic cycle, and the overall climate system.

0.4.2 Government and politics

Delivering on the promise of universal energy access and improved life quality has eluded policymakers and governments over the past seven decades. The affordability of energy services for every global citizen, spanning vastly diverse regions and local contexts, requires the development and massive diffusion of technologies that offer "point-of-use" options combined with new business models. The question becomes, what energy can be redirected to serve as viable sources to meet the human needs in our societies.

Human civilization requires energy to function. Humanity needs energy to power its endeavors into the future. Energy fuels bodies and enables technology. Insufficient energy necessitates an inability to sustain people at the desired technological level. A sufficient energy supply means that the expected number of people will live a full life at their desired technological level. Although fossil energy sources are still plentiful in the world, key technologies and the increasing demand for ecological environmental protection drive the inevitable transition from petroleum to new energy sources.

0.4.3 Economics

The affordability of energy services for every global citizen, spanning vastly diverse regions and local contexts, requires the development and massive diffusion of technologies that offer "point-of-use" options combined with new business models. Social innovations and flexible governance approaches will also need to be integrated with technological advances. The scope and scale of developmental change span large-scale grid systems to decentralized distributed resources at community levels to the households. It will be important to embed low-cost, high-performance, next-generation technological solutions in the field. To meet the needs of those at the base of the economic and social pyramid, the dual challenges of economic development and transition to a low-carbon energy future make clean energy access the quintessential challenge of the 21st century.

0.5 Universal energy into green sustainable energy

The current energy crisis is reshaping previously well-established demand trends. Renewable energy and sustainable development are widely discussed and highly debated topics. The current and majority opinion is that for sustainable development renewable energy is a necessity and plenty of it is available, which can be harvested economically and in environmentally friendly way [49]. Energy harvesting (EH, also known as power harvesting, energy scavenging, or ambient power) is the process by which energy is derived from external sources (e.g., solar power, thermal energy, wind energy, salinity gradients, and kinetic energy, also known as ambient energy), and then stored for use later.

Energy harvesters usually provide a very small amount of power for low-energy electronics. Harvesting sustainable energy from the sun and cold space to uninterruptedly generate green electricity provides a potential alternative way to solve the unfolding energy crisis and environmental issues. This emerging energy technology is seeing important progress and further advances remain challenging. The recent advances in energy harvesters (solar absorbers or/and radiative coolers) integrated with energy converters (thermoelectric generators) provide great prospects for uninterrupted energy harvest and electricity generation in extensive fields [50].

REFERENCES

1. Zohuri, B. and P. McDaniel, Fundamentals of first law energy analysis, in *Encyclopedia of Energy Storage: Volume 1–4*. 2022, Elsevier. p. 5–98.
2. Uzan, J.P., The Big-Bang theory: Construction, evolution and status, in *Progress in Mathematical Physics*. 2021, Birkhauser. p. 1–72.
3. Cork, S., *et al.*, Exploring alternative futures in the anthropocene. *Annual Review of Environment and Resources*, 2023. **48**: p. 25–54.
4. Rogers, N., The Anthropocene judgments project: Futureproofing the common law,in *The Anthropocene Judgments Project: Futureproofing the Common Law*. 2023: Taylor and Francis. p. 1–309.
5. White, P.J., *et al.*, Agency in the anthropocene: Education for planetary health. *The Lancet Planetary Health*, 2024. **8**(2): p. e117–e123.
6. Nelson, W.M., *Sustainable Agricultural Chemistry in the 21st Century: Green Chemistry Nexus*. 2023: CRC Press. p. 1–294.
7. Syvitski, J., *et al.*, Extraordinary human energy consumption and resultant geological impacts beginning around 1950 CE initiated the proposed Anthropocene Epoch. *Communications Earth and Environment*, 2020. **1**(1).
8. Zanna, L., *et al.*, Global reconstruction of historical ocean heat storage and transport. *Proceedings of the National Academy of Sciences of the United States of America*, 2019. **116**(4): p. 1126–1131.
9. Baltay, C., The accelerating universe and dark energy. *International Journal of Modern Physics D*, 2014. **23**(6).
10. Taler, D., Mass, momentum and energy conservation equations, in *Studies in Systems, Decision and Control*. 2019, Springer International Publishing. p. 9–46.
11. Benitez, F., D. Romero-Maltrana, and P. Razeto-Barry, (Re)interpreting E = mc². *Foundations of Physics*, 2022. **52**(1).
12. Zloshchastiev, K.G. Is sustainability of light-harvesting and waveguiding systems a quantum phenomenon? in *Journal of Physics: Conference Series*. 2019. Institute of Physics Publishing.
13. Di Corpo, U. and A. Vannini. Syntropy and sustainability. in *58th Annual Meeting of the International Society for the Systems Sciences, ISSS 2014*. 2014. International Society for the Systems Sciences (ISSS).
14. Gribov, L.A., V.I. Baranov, and I.V. Mikhailov, Some basic statements of the general theory of the universe evolution at the first stages of life. *Geochemistry International*, 2021. **59**(11): p. 1106–1112.
15. He, D., D. Gao, and Q.Y. Cai, Spontaneous creation of the universe from nothing. *Physical Review D - Particles, Fields, Gravitation and Cosmology*, 2014. **89**(8).
16. Driessen, A., The quest for truth of Stephen Hawking. *Scientia et Fides*, 2021. **9**(1): p. 47–61.

17. Adams, F.C., The degree of fine-tuning in our universe: And others. *Physics Reports*, 2019. **807**: p. 1–111.
18. Hay, F.J. and N. Ianno, The development of solar energy generation technologies and global production capabilities, in *Green Energy to Sustainability: Strategies for Global Industries*. 2020, Wiley Blackwell. p. 77–84.
19. Mareschal, J.C. and C. Jaupart, Energy Budget of the Earth. *Encyclopedia of Earth Sciences Series*, 2020. **Part F4**.
20. Tandon, S.K. and R. Sinha, Geology of large river systems, in *Large Rivers: Geomorphology and Management*. 2022, Wiley Blackwell. p. 7–41.
21. Liang, S., *et al.*, Remote sensing of earth's energy budget: Synthesis and review. *International Journal of Digital Earth*, 2019. **12**(7): p. 737–780.
22. Loeb, N.G., W. Su, and S. Kato, Understanding climate feedbacks and sensitivity using observations of earth's energy budget. *Current Climate Change Reports*, 2016. **2**(4): p. 170–178.
23. Buscheck, T.A., J.M. Bielicki, and J.B. Randolph. CO2 earth storage: Enhanced geothermal energy and water recovery and energy storage. in *Energy Procedia*. 2017. Elsevier Ltd.
24. Ramsey, K.M. and J. Bass, Circadian clocks in fuel harvesting and energy homeostasis. *Cold Spring Harbor Symposia on Quantitative Biology*, 2011. **76**: p. 63–72.
25. Zhang, H., *et al.*, Earth's energy budget, climate feedbacks, and climate sensitivity. *Climate Change Research*, 2021. **17**(6): p. 691–698.
26. Swift, L., A new radiativemodel derived from solar insolation, albedo, and bulk atmospheric emissivity: Application to Earth and other planets. *Climate*, 2018. **6**(2).
27. Wang, K., *et al.*, A multi-model assessment for the 2006 and 2010 simulations under the Air Quality Model Evaluation International Initiative (AQMEII) Phase 2 over North America: Part II. Evaluation of column variable predictions using satellite data. *Atmospheric Environment*, 2015. **115**: p. 587–603.
28. Stamatis, M., *et al.*, Interdecadal changes of the MERRA-2 incoming surface solar radiation (SSR) and evaluation against GEBA & BSRN stations. *Applied Sciences (Switzerland)*, 2022. **12**(19).
29. Fang, L., Dynamics of renewable energy index in G20 countries: influence of green financing. *Environmental Science and Pollution Research*, 2023. **30**(23): p. 63811–63824.
30. Goel, M., V.S. Verma, and N.G. Tripathi, Sun: Unlimited energy resource on earth, in *Green Energy and Technology*. 2022, Springer Science and Business Media Deutschland GmbH. p. 15–26.
31. Kennedy, I.R. and M. Hodzic, Applying the action principle of classical mechanics to the thermodynamics of the troposphere. *Applied Mechanics*, 2023. **4**(2): p. 729–751.
32. Rasmussen, K.L., *et al.*, Changes in the convective population and thermodynamic environments in convection-permitting regional climate simulations over the United States. *Climate Dynamics*, 2020. **55**(1–2): p. 383–408.
33. Von Schuckmann, K., *et al.*, An imperative to monitor Earth's energy imbalance. *Nature Climate Change*, 2016. **6**(2): p. 138–144.

34. Zhang, A., *et al.*, Forecasting the progression of human civilization on the Kardashev Scale through 2060 with a machine learning approach. *Scientific Reports*, 2023. 13(1).

35. Dissanayake, H., *et al.*, Nexus between carbon emissions, energy consumption, and economic growth: Evidence from global economies. *PLoS ONE*, 2023. 18(6 June).

36. Mischos, S., E. Dalagdi, and D. Vrakas, Intelligent energy management systems: a review. *Artificial Intelligence Review*, 2023. 56(10): p. 11635–11674.

37. Kellogg, K.A. and N.L. Cumbre-Gibbs, The impact of state level residential building code stringency on energy consumption in the United States. *Energy and Buildings*, 2023. 278.

38. Iyer, A.V., N.D. Rao, and E.G. Hertwich, Review of Urban building types and their energy use and carbon emissions in life-cycle analyses from low- and middle-income countries. *Environmental Science and Technology*, 2023. 57(26): p. 9445–9458.

39. Mukherjee, P.K., *et al.*, Socio-economic sustainability with circular economy: An alternative approach. *Science of the Total Environment*, 2023. 904.

40. Adekoya, O.B., T.P. Ogunnusi, and J.A. Oliyide, Sector-by-sector non-renewable energy consumption shocks and manufacturing performance in the U.S.: Analysis of the asymmetric issue with nonlinear ARDL and the role of structural breaks. *Energy*, 2021. 222.

41. EIA, U.S.E.I.A., *Annual Energy Outlook 2022*. 2022.

42. Administration, U.S.E.I., *Monthly Energy Review*. 2023.

43. Wang, Y. and Q. Miao, The impact of the corporate average fuel economy standards on technological changes in automobile fuel efficiency. *Resource and Energy Economics*, 2021. 63.

44. Agency, I.E. 2023. Available from: https://www.iea.org › energy-system › industry.

45. Boretti, A., Supply of abundant and low-cost total primary energy to a growing world needs nuclear energy and hydrogen energy storage. *International Journal of Hydrogen Energy*, 2023. 48(5): p. 1649–1650.

46. Vehmas, J., J. Kaivo-oja, and J. Luukkanen, Energy efficiency as a driver of total primary energy supply in the EU-28 countries: Incremental decomposition analysis. *Heliyon*, 2018. 4(10).

47. Department, S.R. *Global Electricity Mix 2022, by Energy Source*. Sep 7, 2023. Available from: https://www.statista.com/statistics/269811/world-electricity-production-by-energy-source/.

48. Tvaronavičienė, M. and B. Ślusarczyk, Energy transformation towards sustainability. *Energy Transformation Towards Sustainability*. 2019: Elsevier. p. 1–333.

49. Ray, P., Renewable energy and sustainability. *Clean Technologies and Environmental Policy*, 2019. 21(8): p. 1517–1533.

50. Zhang, S., *et al.*, An emerging energy technology: Self-uninterrupted electricity power harvesting from the sun and cold space. *Advanced Energy Materials*, 2023. 13(19).

Unit I

Green Sustainable Energy

We exist in a universe of energy, but it is not always accessible. Green sustainable energy (GSE) is the best option due to climate change. GSE is energy that is clean, has minimal/no effects on the environment, and comes from renewable resources. GSE adoption is inevitable as governments are trying to identify viable solutions to the energy crisis and reduce dependence on fossil fuels. The large-scale transition to GSE will contribute to sustainable development.

Sustainability is now a moral imperative that is directing our energy choices. It will require metrics and indicators to be established. A PESTEL (Political, Economic, Social, Technological, Legal, and Environmental) analysis reinforces the desirability of GSE. PESTEL coupled with Life cycle and multi-criteria decision analyses properly guide GSE. Energy in the 21st century must also be resilient. To achieve an energy system that is both sustainable and resilient, our world must transition to new energy sources, green harvesting of energy, and renewable practices.

The transition to green sustainable energy aligns with the growth of a circular economy; both are fundamental to a sustainable future. The "energy-circularity nexus" provides a holistic approach to energy development, harvesting, and use. As governments, industries, and societies grapple with the complexities of global energy demand, security, economics, and sustainability, comprehensive insights and strategic guidance are needed. Green growth (growth of economies using modern, environmentally sound, and resilient technologies) must include the use of GSE. This is vital since energy is an indispensable part of modern society and serves as one of the most important indicators of socio-economic development of civilization.

DOI: 10.1201/9781003407447-2

Chapter 1

Necessity of green sustainable energy

Energy is essential for development, and sustainable energy is essential for sustainable development.

(Tim Wirth, former US Senator)

1.1 INTRODUCTION

The global need for energy and associated services is vital to meet the demands of human, social, and economic progress, welfare, and health [1–3]. Energy needed to ensure a healthy sustainable life is inseparable from our global environment. Natural energy resources play a key role in the development of human civilization; however, their supply and use have not been managed well. Therefore, one of the most important challenges for the world today is to identify and harvest green sustainable energy (see Figure 1.1) [4].

The energy system (extraction and use) faces numerous challenges, most notably high consumption levels, lack of energy access, environmental concerns like climate change and air pollution, energy security concerns, and the need for a long-term plan for management. A fundamental transformation of the global energy system is needed. Recent publications show that such transformational pathways are achievable in technological and economic terms but constitute formidable governance challenges at larger scales. This transformation would need to incorporate several key components, including developing an integrated approach and maintaining focus on energy efficiency and the scale-up of investments in R&D [5].

Meeting global energy needs in a sustainable and environmentally responsible way is a challenge. The use of energy based on fossil fuels has enabled great advances and an increase in the standard of living, but they are not in infinite supply and have brought us to the brink of an environmental catastrophe. As a society, we will need to develop strategies that integrate green and sustainable energy sources. Global energy policies must be scientifically and technically sound and must also be acceptable. This energy challenge is

DOI: 10.1201/9781003407447-3

21

Figure 1.1 Sustainable energy.
Source: Shutterstock Photo ID: 751358878.

the type of problem that will require decisive scientific and political leadership. This will require leaders to articulate a global energy policy capable of meeting such a challenge [6]. Perfect solutions will be hard to come by, due not only to drastic differences in political and public support for sustainable energy throughout the world but also the extensive knowledge required to address the many challenges associated with the global energy landscape.

More studies are appearing to focus on the inner process of the environmental dimension of the global energy policy (i.e., the axis related to the environmental policy and concerns such as the development of renewable energy, energy efficiency and savings, or reduction of greenhouse gas (GHG) emissions). Environmental concerns regarding energy consumption are a significant driver of global energy policy. The "green" dimension of any energy policy is more important than in earlier periods of history. The concern for green has an overwhelming relevance to climate change. Importantly, it must drive a perspective on the future of global energy policies taking into consideration the current circumstances [7].

1.2 POSSIBILITY OF GREEN SUSTAINABLE ENERGY

The current increasing demand for renewable-energy-based efficient systems is a signal that sustainable energy technologies for harvesting, energy storage, and conversion to reduce the use of pollution-based nonrenewable

Figure 1.2 Green sustainable energy (GSE).

Source: Shutterstock Photo ID: 2483971489

energy systems are needed. The global renewable energy landscape is changing rapidly. Green energies can reduce greenhouse gas emissions, diversify the energy supply, and lower dependence on volatile and uncertain fossil fuel markets. Governments must try to identify viable solutions to the energy crisis and reduce dependence on fossil fuels.

Green sustainable energy (GSE) is highly desirable but needs to be developed further [8] (see Figure 1.2). Sustainable energy is produced from resources that can meet current demands without jeopardizing the energy needs or climate of future generations. Renewable energy, on the other hand, comes from sources that naturally sustain or replenish themselves over time. Finally, green energy is any energy type that is generated from natural resources and does not harm the environment, such as sunlight, wind, or water. GSE combines all these attributes. By taking advantage of the Earth's innate sustainable environment, GSE sources will theoretically be able to supply our energy needs indefinitely, while protecting/preserving the environment.

Renewable energy and sustainable development must be components of GSE. It is believed that for sustainable development, renewable energy is a necessity, which can be harvested economically and environmentally friendly. It is held that fossil fuels are far from exhausted and can be used with clean technologies, which are already developed, while technical problems for renewable energy are far from solved and they are very often more damaging to the environment and society than envisaged. The problems of global warming and carbon dioxide build-up are also inseparable from

SGE. A concept of sustainability steady state for energy is the universal state presented in the prolog ("A Universe of Energy"). The correlation between carbon dioxide build-up and global warming necessitates that renewable energy will have to be adopted. This will have to be aided by increased amounts of indirect solar energy like wind energy and particularly bioenergy. Renewable sources, while having orders of magnitude greater energy content than human society may consume, are not particularly easy to harness, allowing only a small part to be finally harvestable. There are tough technical, environmental, and societal problems, all quite significant, that have to be solved. Thus, in the long run, GSE will become inevitable, but even this will require a great deal of effort and planning and will not come easily [9].

1.2.1 Green energy and sustainability

Despite increasing desires to move to green economies, much remains to be done globally to make the energy transition a reality. Beyond the desirability of this type of energy, there will be political, economic, and technical challenges (discussed in Chapter 2, "PESTEL analyses of GSE"). The current global circumstances underline the urgency of addressing energy development and consumption and its environmental implications. There is a critical role of practical action in fostering energy sustainability and environmental preservation [10]. For example, with the setting of goals, the European Union member states have committed to a wide neutrality target by establishing an increase in the total share of energy from renewable sources to 55% (by 2030) and, at the same time, reducing the net greenhouse gas effect emissions by at least 55% until 2030 to reach the neutrality target by 2050 [1].

Nations have relied upon fossil fuels for primary energy and this will remain for a few more decades. Other energy sources may be more enduring, but they also have serious disadvantages. Power from natural resources was readily available and plentiful. Coal was abundant (though known supplies are dwindling [11]), and there is concern about despoliation in mining it and pollution in burning it. Nuclear power has been developed with remarkable timeliness, but is not universally welcomed, construction of the plant is energy-intensive, and there is concern about the disposal of its long-lived active wastes. Barrels of oil, lumps of coal, and even uranium come from nature but the possibilities of almost limitless power from the atmosphere and the oceans seem to have special attraction. The wind machine provided an early way of developing motive power. The speculation that we can draw upon universal energy as fuel appears to be more attractive and to be worth reinvestigation. In addition to direct solar energy, the atmosphere and the oceans are energy sources and can result in wind power, wave power, and tidal power from ocean thermal gradients [12].

Table 1.1 Examples of renewable energy sources

Category	Sources	Examples/uses
Biomass	Organic material that is burned or converted to liquid or gaseous form	Ethanol and biodiesel
Geothermal	Heat produced by decaying radioactive particles found deep within the Earth	Direct heat source or to generate electricity
Hydropower	Flowing water	Hydropower was the largest source of renewable electricity until 2019
Solar	Requires a large surface area and consistent sunlight	Harness the sun's potential
Wind	Utilizes turbines to convert the wind's kinetic energy into mechanical energy	Stored or used to accomplish a task like grinding grain

1.2.1.1 Renewable energy sources

Produced from existing resources that naturally sustain or replenish themselves over time, renewable energy is a much more sustainable solution than our current top energy sources. Renewable energy sources are increasingly cost-efficient, and their impact on the environment is far less severe than fossil fuels. By taking advantage of the Earth's ability to grow and recycle organisms, renewable power sources will theoretically be able to supply our energy needs indefinitely. Renewable energy is defined by the time it takes to replenish the primary energy resource, compared to the rate at which energy is used. Traditional resources like coal and oil, which take millions of years to form, are not considered renewable. On the other hand, solar power can always be replenished, even though conditions are not always optimal for maximizing production.

Examples of renewable energy sources are included in Table 1.1. All sources listed will require green ways to store the energy produced. This will be elaborated in Chapter 8, "GSE harnessing, harvesting, and storage."

1.2.1.2 Sustainable energy sources

Sustainable energy must be obtained from resources that can be extracted with current technologies without jeopardizing the energy needs or climate of future generations. The most useful sources of sustainable energy, including wind, solar, and hydropower, are also renewable. Biomass is an energy source that is not entirely sustainable, for example. Biofuel is a form of renewable energy, but its consumption emits climate-affecting greenhouse gasses, and growing the original plant product uses up other environmental resources. However, biofuel is a major component of the green revolution.

The key challenge with biofuel, as with the other sources, is finding ways to maximize energy output while minimizing the impact of sourcing biomass and burning the fuel. Even with resources that are both renewable and sustainable, like wind and solar power, an important question remains: Is sustainable energy the solution to our energy and climate needs?

The transition to green sustainable energy is filled with nuanced options and not simple. All sustainable solutions cannot be used in every situation. Their efficiency and/or effectiveness depends on factors such as climate and location, and once the energy is generated, collected, and stored, it must then be distributed. For instance, wind is produced by temperature changes in the air, which aren't consistent across the planet. In the United States, this means that the best place to put wind farms is in the Midwest. However, storage and distribution present challenges. Disparities in regulations and target goals can create a problem where the best place to produce energy may not have the public interest or infrastructure necessary to support it. For example, a windy state may struggle to pass legislation for financing the construction of turbines, while its neighbor may be eager for a nearby source of clean energy.

1.2.1.3 Green energy sources

Green energy is energy from renewable resources that are naturally replenished on a human timescale. This energy (harvesting, storage, and use) has minimal or no impact on our environment. When green energy and renewable energy, such as solar and wind, are combined, the result is the ideal clean energy mix [13]. This is GSE. To be truly green energy, a resource cannot produce pollution, such as is found with fossil fuels, and they are not sustainable. This means that not all sources used by the renewable energy industry are green. For example, power generation that burns organic material from sustainable forests may be renewable, but it is not necessarily green, due to the CO_2 produced by the burning process itself. Green energy sources are usually naturally replenished, as opposed to fossil fuel sources like natural gas or coal, which can take millions of years to develop. Green sources also often avoid mining or drilling operations that can be damaging to ecosystems.

Green energy sources are essential for long-term efforts to mitigate climate change and will play an important role in improving energy security and accessibility. The efforts of every country to strengthen the energy sector through the development of green energies will reduce geopolitical risks and disproportionate external costs for society. The large-scale use of GSE will contribute to sustainable development. The technical challenges, key factors, and future research directions of GSE technologies and systems show that that effective, intelligent, and innovative integration of them with other technologies will encourage more efficient use of resources and enhance energy system flexibility, resilience, and energy security [14].

The global capacity to deploy green renewable energy technology or participate in climate change mitigation is geographically variable and no single solution is universally viable. Green energy must go beyond decarbonization and be more comprehensive and globally inclusive. Strategies pivotal to this approach involve understanding the role of carbon-based energy systems in the transition, amplifying renewable resources, augmenting cross-sector energy efficiency, implementing effective carbon markets, and integrating nature-based as well as carbon removal technologies. Moreover, it is imperative to implement all-cost and all-benefit monitoring and evaluation systems to optimize existing decarbonization methods systematically. Addressing societal apprehensions requires a focus on pragmatic and fair outcomes, geopolitical stability, market impacts, developmental objectives, effective public engagement, and recognition of the role of enterprises. Policymakers are critical in fostering global synergy by implementing policies that encourage international collaboration, investment, enterprise engagement, institution fortification, and cross-sector policy integration [15].

The effectiveness of green energies will be measured by how they can contribute to sustainable development. According to international statistics, more than 90% of the governments of many countries are making investments to efficiently capitalize on green energy sources and to design new models of sustainable economic and social development, in order to lower pollution levels, reduce the dependence on fossil fuels, and limit the climate change impact [1]. The transition to green energy is aimed at mitigating the impact of climate change and thus is necessary to achieve sustainability. Yet, the current emphasis on "green" is narrowly centered around decarbonization, or CO_2 reduction, often disregarding the roles of other gases, such as sulfur hexafluoride (SF_6) and PCF-14 (CF_4), which have a respective 24,300- and 7380-times higher global warming potential than CO_2 on a time horizon of a century [15]. Therefore "green" has much to contribute to sustainability.

1.2.2 Green sustainable energy (GSE)

The link between green sustainable energy and improving environmental quality standards has significant repercussions for the long-term viability of the ecosystem. There is a nexus between sustainable energy, green energy, and environmental health. GSE has the potential to be a key tool in the achievement of environmental sustainability. Energy efficiency, globalization, and greener energy adoption are helpful for the reduction of GHG emissions in developed countries. Policies for greener energies can ensure sustainable development [16].

Some of the actual challenges regarding green sustainable energy systems come from their specific source [17], meaning the nature of the source, which can make the type of energy intermittent, depending on the weather conditions, or the type of technology used, the storage capacity, the trained support needed where infrastructure is installed, and lack of maintenance systems [18]. Green sustainable energy sources are clean sources of energy, depending on the technology and environmental impact, including sun, wind, geothermal, hydroelectricity, tidal and ocean energy, and biomass [19].

The global economy is at the early stage of a green transformation. This transformation must further be marked by the commitment to sustainability. It is clear that multinational enterprises (MNEs) should seek to enhance their capabilities for sustainable innovation and many have started to globalize their green efforts. Green Foreign Direct Investments (FDIs) enhance the overall orientation to sustainability of MNEs. Secondly, the MNEs extend their innovative capabilities toward more sustainability-oriented direction and strengthen their innovation activities related to green technologies. Finally, the globalization process mode matters: in the long run, green economics result in new subsidiaries contributing more to innovativeness and greening [20].

1.2.2.1 Green sustainable fuels

Green sustainable fuels are fuels produced from renewable resources that have minimal or no effects on the environment and do not affect future generations. Examples include biofuels (e.g., vegetable oil used as fuel, ethanol, methanol from clean energy and carbon dioxide or biomass, and biodiesel), hydrogen fuel (when produced with renewable processes), and fully synthetic fuel (also known as electrofuel) produced from ambient carbon dioxide and water. These fuels can include fuels that are synthesized from GSE sources, such as wind and solar. These fuels are also valuable due to their sustainability and low contributions to the carbon cycle and they produce lower amounts of greenhouse gases.

Green sustainable energy sources will play a vital role in global development. As an interim measure, renewable biofuels for transport represent a key source of diversification from petroleum products. Biofuels from grain and beet in temperate regions have a part to play, but they are relatively expensive and their energy efficiency and CO_2 savings benefits are variable. Biofuels from sugarcane and other highly productive tropical crops are much more competitive and beneficial. But all first-generation biofuels ultimately compete with food production for land, water, and other resources. Biofuels made from sustainable sources are essential for the global economy's long-term viability and the reduction of greenhouse gas emissions. They are graded transition from fossil fuel dependence. Clean, renewable, and sustainable must be the watchwords for future energy strategy. Alcohol fuels are also examples of transition to green fuel usage in relation to climate change mitigation and clean fuel technology [21, 22].

1.2.2.2 Green sustainable power

Green sustainable power is often synonymous with sustainable energy. This encompasses most energy uses beyond fuels. It encompasses GSE resources and technologies that provide the greatest environmental benefit. Within the U.S. green power is defined as electricity produced from solar, wind, geothermal, biogas, eligible biomass, and low-impact small hydro-electric sources. Green power should have zero-emissions profile and carbon footprint reduction benefits. As was the case with fuel, these sources do produce CO_2.

Green sustainable power (GSP) can impact the environment, leading to lower sustainability. GSP comes from resources that rely on fuel sources replenished over short periods, such as the sun, wind, moving water, organic plant and waste material (eligible biomass), and the Earth's heat (geothermal). GSP should not have environmental impacts. For instance, large hydroelectric resources may have trade-offs related to fisheries and land use. In contrast, conventional power relies on the combustion of fossil fuels (coal, natural gas, and oil) or the nuclear fission of uranium. Fossil fuels come with environmental costs from mining, drilling, and emissions, while nuclear power requires mining, extraction, and long-term radioactive waste storage.

1.3 GLOBALIZATION

The global awareness of the world's degradation has brought us to a tip-ping point. The degradation has resulted in a tremendous increase in species extinction, increased deforestation rates, and increased proximity to the 2-degree critical temperature thresholds. Globalization is leading the modern world toward green sources of energy. GSE must meet current needs and have a lower environmental impact compared to conventional energy technology. The balance among the effects of technology innovation, international poverty, and energy consumption has considerable effects on the load capacity factor (LCF). Investments in research and development to foster technological innovation across various sectors while enforcing environmental compliance to minimize the negative externalities associated with nonrenewable activities are necessary actions that must be taken to assure GSE globalization (see Figure 1.3) [23].

Given improving sustainable green energy options, detailed green energy planning for sustainable development must be employed to provide the most suitable common strategy for energy management [24]. Answers to the following questions would be helpful:

1. Which approaches can be applied in a particular region but may be applied globally?

Figure 1.3 Globalization of energy.

Source: Shutterstock Photo ID: 2297429939.

2. Which evaluating criteria are necessary for energy planning for sustainable development?
3. Are there inadequacies or weaknesses in the approaches?
4. Will GSE meet current and anticipated energy demands?

1.3.1 Main categories for sources of GSE

Green sustainable energy is fundamental for achieving sustainable development. Not all the resources employed to produce energy by the renewable energy sector are GSE. The future looks promising for green energy sources, which are taking on an increasingly important role, especially as governments are trying to identify viable solutions to the energy crisis and reduce dependence on fossil fuels. Worldwide, there is a growing interest in and support for green energy sources, a factor that could help accelerate the current energy transition. Despite these positive developments, much remains to be done globally. In 2019 at the global level, low-carbon energy sources, including nuclear power and renewable energy, accounted for 15.7% of primary energy (solar, wind, hydropower, bioenergy, geothermal, and wave and tidal), while in 2021, for the European Union, the share of energy from renewable sources reached 22%. According to international statistics, more than 90% of the governments of many countries are making investments to efficiently capitalize on green energy sources and to design new models of sustainable economic and social development [1]. In the following, we list some sources of GSE and will discuss them further in the subsequent chapters.

- Solar energies [25]
- Food waste [26]

- Thermal energy storage [27]
- Biomass [28, 29]
- Ammonia
- Hydrogen [30]

1.3.2 GSE through machine learning (ML) and artificial intelligence (AI)

The global condition of GSE adoption must include their benefits and limitations in meeting energy needs. ML and AI can assist in promoting the growth of solar and wind energy, global trends in the usage of GSE, and upcoming technologies, including floating solar and vertical-axis wind turbines. The importance of smart grid technology and energy storage alternatives (see Chapter 8, "GSE harnessing, harvesting, and storage") for enhancing the effectiveness and dependability of renewable energy is critical in this effort. ML/AI applications for solar and wind energy generation are vital for sustainable energy production. Machine learning can help in design, optimization, cost reduction, and, most importantly, in improving the efficacy of solar and wind energy, including advancing energy storage. Ultimately, this technology has shown the great potential of GSE in meeting global energy demands and sustainable goals [31, 32].

1.4 GSE AND ENERGY DEMANDS IN THE 21ST CENTURY

In the 21st century, green sustainable energy offers alternatives for continued global development. GSE sources are a strategic option for sustainable development [33]. The identification and use of GSE sources should be a fundamental element of sustainable development. It is essential to place GSE at the center of the sustainable development paradigm as a means of meeting the international agreements and SDG 7 challenges [34].

Highlighting the components and main advantages of SGE will allow both consumers and policymakers to realize the importance of steering the energy system in the direction of a long-term sustainable development. It is valuable to identify which strategies and initiatives are ineffective for achieving the various goals related to the economy, energy, sustainability, and climate change. The adoption of GSE must be achieved through the promotion of international collaboration in science, science–policy interaction, and information transfer.

The following are areas of importance that will clarify the energy demand that will influence the growth of GSE:

- Education
- Sustainability boundaries

- Storage
- Metrics
- Global and local impact
- Cost

Through collaborative efforts from the public and private sectors, increasing the use of GSE sources and techniques can achieve better outcomes for the environment, energy, security, and sustainable development.

1.4.1 Education

Efforts to aggressively decarbonize the energy economy and develop new technologies, especially for the generation of electrical energy that is environmentally clean, are exemplified by the recent climate change agreement in Paris [35]. The challenge must include a deliberate approach to research and education of the next generation of scientists and engineers as well as informing the general public. This will necessitate the introduction of new and comprehensive education programs on sustainable energy technologies for universities and, possibly, high schools. The new programs should provide in-depth knowledge in the development of new materials for more efficient energy conversion systems and devices. The significance of clean and efficient energy systems (development, harvesting, storage, and use) indicates the need for a comprehensive education program [36].

1.4.2 Sustainability boundaries

Sustainability boundaries, or guardrails, in all global energy scenarios must be developed. Environmental planetary boundaries mark out the safe operation space for human activities. The idea of a planetary boundary framework needs to be fully incorporated into green sustainable energy scenario modeling, going beyond CO_2 emission mitigation. There needs to be more consideration of biochemical flows, land use change, biodiversity, ocean, and climate systems. Concurrently, social and economic dimensions, such as limiting air pollution, providing universal access to modern energy services, and improving energy efficiency and energy services, are emerging as new paradigms that should be incorporated into energy scenario modeling frameworks. Moreover, GSE will involve ethical choices, such as current and future generations' access to preserved ecosystems, aversion of energy resource risks, preventing resource use conflicts, and negative impacts on human lives from energy extraction and use [37, 38].

1.4.3 Storage

A heightened interest in clean energy sources will motivate many countries to implement GSE technologies. These efforts are critical for moving the world toward decarbonization and decreasing emissions within the transportation and power industries. A challenge with renewable sources (particularly solar and wind) is the inherent variability of energy generation due to climate and geographic influences. Consequently, there is increased requirement in addressing energy storage technologies to better capitalize on excess power from high-production periods for use in times of low generation. There are three areas of potential policy improvements for industry:

- Implementation of a policy framework for states or regions to produce ambitious energy storage technologies and distribution metrics.
- Incentivizing the federal investment tax credit for energy storage technologies to become universally available.
- Improvement of sustainable battery technologies.

It is imperative for world governments to create informed energy storage goals in tandem with current renewable energy metrics. This will allow for the continued deployment of batteries, incorporating sustainable practices to mitigate material and environmental challenges posed by them, including recycling practices and alternative chemistry utilization [39].

1.4.4 Metrics

The value of GSE will be felt in its contributions to the United Nations (UN) Sustainable Development Goals (SDGs) [40]. In order to achieve this, credible objective metrics to gauge progress will strengthen a case for continued financial investing. A set of science-based metrics allows corporations and interested investors to meaningfully choose to use SGEs in locations around the world where they can make the greatest positive impact. In particular, this aligns with a new framework and methodology that meets the deployment of Affordable and Clean Energy (SDG 7) to explicitly measure advances [41, 42]. The methodology to better measure progress toward climate, energy, and health-related SDGs in financial investing and other applications will greatly assist the achievement of sustainability and adoption of GSE.

Green chemistry is one of the most influential concepts widely adopted by chemical industries to achieve sustainability. It has led to a better understanding of sustainable ways to improve chemical processes using green metrics. Over the last two decades, studies on various metrics and their correlation were formulated to evaluate existing chemical processes. This may also provide the metrics to measure GSE. Green metrics allow the quantitative understanding of energy and cost-efficiencies that provide strategic planning called "green-by-design." Improving green metrics for assessing GSE for their greener qualities and sustainability makes them broadly applicable [43]. This may now be combined with the use of exergy as an indicator for energy sustainability, and we will now have a global energy sustainability indicator. Exergy is a thermodynamic property that links the first and the second thermodynamic principles as well as connects the system to the environment where it belongs. Since the first principle of thermodynamics measures the quantity of energy and the second measures irreversibilities, i.e., quality of energy, having a single thermodynamic indicator that can incorporate both principles at the same time means a great advance in energy sustainability studies. Using exergy provides a valuable and economic analysis for SGE [44].

1.4.5 Global and local impact

Using exergy as an indicator for energy sustainability studies, it can also be used as a global energy sustainability indicator [45]. The measure of the true value of GSE will be found in both its global and local impacts. Evaluating global and local impacts will yield several conclusions:

- Sustainable green development in any region will vary.
- The spatial characteristics of sustainable green development efficiency differ and will vary according to the indices used. However, when taken in aggregate, the results will show improvements in overall green sustainability.
- Technological progress, increasing communication, and urbanization levels are positively correlated with the green development efficiency. Industrial structure, financial development, energy structure, and environmental regulation show mixed impact [46].

Through effective supply chain management, responsible sourcing, green logistics, and waste reduction, economic resilience can be bolstered. Also, glocalization, which seeks to reconcile global and local viewpoints, plays a vital role in ensuring economic stability [47]. The importance of these factors in the adoption and use of GSEs cannot be overstated.

1.4.6 Cost

Solar and wind power will someday become the cheapest way to generate electricity, allowing the world to shift away from fossil fuels. Once this occurs, there is a way for renewables to eventually account for most of global electricity and energy production, far beyond today's 26% share. As the world transitions to a green sustainable future, there will be felt changes.

There will be a decline in prices, but the rate at which these prices fall is also expected to be rapid. For example, every year for the last decade, electricity from solar and wind has cost less than the previous year. GSE is now approaching parity with the cost of building new coal and nuclear

Trends in cost of energy (2010-2019)

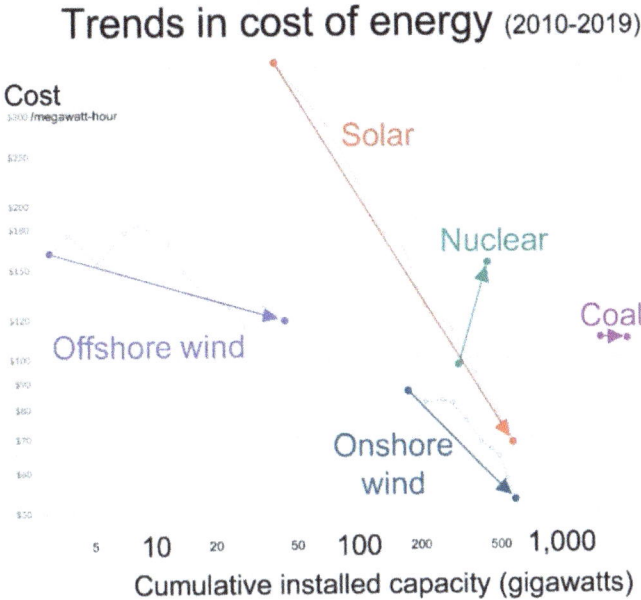

Figure 1.4 Trends in cost of energy. (By Our World in Data - Max Roser – https://ourwo rldindata.org/cheap-renewables-growth; //commons.wikimedia.org/w/index.p hp?curid=106244999).

energy facilities. As existing plants reach retirement ages, GSE technologies will be built and used.

The levelized cost of electricity (LCOE) is a metric that attempts to compare the costs of different methods of electricity generation consistently. Though LCOE is often presented as the minimum constant price at which electricity must be sold to break even over the lifetime of the project, such a cost analysis requires assumptions about the value of various nonfinancial costs (environmental impacts, local availability, others), and is therefore controversial. Roughly calculated, LCOE is the net present value of all costs over the lifetime of the asset divided by an appropriately discounted total of the energy output from the asset over that lifetime [48] (see Figure 1.4).

1.5 LESSONS

There are lessons from the history of technology introductions that should not be forgotten when considering the development and introduction of alternative energy technologies for carbon dioxide emission reductions. Since the growth of the ecological footprint of the human population will increase from 7 billion now to 9 billion people in 2050, there are serious concerns about how to live both more efficiently and with less permanent impacts on the finite world. The future of our climate has prompted actions across the world in mitigation of the emissions of CO_2. An examination of successful and failed introductions of technology over the last 200 years to 80% decarbonize the world economy by 2050 shows that all the actions taken together until now to reduce our emissions of carbon dioxide will not achieve a serious reduction. This means that new technology introductions are needed to be able to meet the huge existing demand. More sophisticated public debate is urgently needed on appropriate actions that

1. consider the full range of threats to humanity, and
2. weigh more carefully both the upsides and downsides of taking any action and of not taking that action [49].

1.5.1 Green sustainable energy in perspective

Conventional energy sources and usage based on nonrenewable fossil fuels have promoted historical contributions to human societies and the advancement of civilization. However, it has led to environmental degradation which the biogeochemical cycles of carbon, nitrogen, and phosphorus cannot remediate. Therefore, it has become necessary to rethink how we derive the energy we need, that is, "the replacement of a philosophy that saw energy generation as an 'end in itself' with one that aligns with natural principles of sustainability and renewability." By utilizing what we have learned about universal energy (see the "Prolog") and involving the techniques and tools of green sciences, we can envision a supply of energy that will be both green and sustainable [50].

In the chapters that follow, available resources and developing technologies will be presented that will show that this is a realizable goal. Utilizing wisdom gleaned from Nature as well as developing guidance from circularity, we as a global society may accomplish sustainable practices in the vital area of energy generation and usage.

REFERENCES

1. Androniceanu, A. and O.M. Sabie, Overview of green energy as a real strategic option for sustainable development. *Energies*, 2022. 15(22).
2. Hassan, Q., *et al.*, The renewable energy role in the global energy Transformations. *Renewable Energy Focus*, 2024. 48.
3. Gielen, D., *et al.*, The role of renewable energy in the global energy transformation. *Energy Strategy Reviews*, 2019. 24: p. 38–50.
4. Jiang, Q., *et al.*, Environment, energy, sustainability: Journal- ES energy & environment. *Engineered Science*, 2018. 3: p. 1–4.

5. van Vuuren, D.P., *et al.*, An energy vision: The transformation towards sustainability-interconnected challenges and solutions. *Current Opinion in Environmental Sustainability*, 2012. 4(1): p. 18–34.

6. Abruña, H.D., Energy in the age of sustainability. *Journal of Chemical Education*, 2013. 90(11): p. 1411–1413.

7. Morata, F. and I. Solorio Sandoval, When 'green' is not always sustainable: The inconvenient truth of the EU energy policy. *Carbon Management*, 2013. 4(5): p. 555–563.

8. Arshad, M.Y., *et al.*, Role of experimental, modeling, and simulation studies of plasma in sustainable green energy. *Sustainability (Switzerland)*, 2023. 15(19).

9. Ray, P., Renewable energy and sustainability. *Clean Technologies and Environmental Policy*, 2019. 21(8): p. 1517–1533.

10. Soni, N., *et al.*, Advancing sustainable energy: Exploring new frontiers and opportunities in the green transition. *Advanced Sustainable Systems*, 2024.

11. Jonsson, F.A., The coal question before jevons. *Historical Journal*, 2020. 63(1): p. 107–126.

12. Omer, A.M., The green revolution: Development of sustainable energy research and applications, in *Advances in Environmental Research*. 2014, Nova Science Publishers, Inc. p. 149–182.

13. Shen, G., *et al.*, Substantial transition to clean household energy mix in rural China. *National Science Review*, 2022. 9(7).

14. Kourougianni, F., *et al.*, A comprehensive review of green hydrogen energy systems. *Renewable Energy*, 2024. **231**.

15. Ghorbani, Y., *et al.*, Embracing a diverse approach to a globally inclusive green energy transition: Moving beyond decarbonisation and recognising realistic carbon reduction strategies. *Journal of Cleaner Production*, 2024. **434**.

16. Pi, K., *et al.*, Sustainable energy efficiency, greener energy and energy-related emissions nexus: Sustainability-related implications for G7 economies. *Geological Journal*, 2024. 59(1): p. 301–312.

17. Kaygusuz, K., Energy for sustainable development: Key issues and challenges. *Energy Sources, Part B: Economics, Planning and Policy*, 2007. 2(1): p. 73–83.

18. Wang, B., L. Li, and X. Jiang, Sustainable development of energy systems and climate systems: Key issues and perspectives. *Energy Engineering: Journal of the Association of Energy Engineering*, 2023. 120(8): p. 1763–1773.

19. Panwar, N.L., S.C. Kaushik, and S. Kothari, Role of renewable energy sources in environmental protection: A review. *Renewable and Sustainable Energy Reviews*, 2011. 15(3): p. 1513–1524.

20. Amendolagine, V., R. Lema, and R. Rabellotti, Green foreign direct investments and the deepening of capabilities for sustainable innovation in multinationals: Insights from renewable energy. *Journal of Cleaner Production*, 2021. **310**.

21. Bilgili, F. and H.H. Bağlıtaş, The dynamic analysis of renewable energy's contribution to the dimensions of sustainable development and energy security. *Environmental Science and Pollution Research*, 2022. 29(50): p. 75730–75743.

22. Anekwe, I.M.S., *et al.*, Sustainable fuels: Lower alcohols perspective. *Environmental Progress and Sustainable Energy*, 2023.
23. Ali, E.B., *et al.*, Load capacity factor and carbon emissions: Assessing environmental quality among MINT nations through technology, debt, and green energy. *Journal of Cleaner Production*, 2023. **428**.
24. Bhowmik, C., *et al.*, Optimal green energy planning for sustainable development: A review. *Renewable and Sustainable Energy Reviews*, 2017. **71**: p. 796–813.
25. Abdalla, A.N., *et al.*, Socio-economic impacts of solar energy technologies for sustainable green energy: A review. *Environment, Development and Sustainability*, 2022.
26. Ma, Y. and Y. Liu, Turning food waste to energy and resources towards a great environmental and economic sustainability: An innovative integrated biological approach. *Biotechnology Advances*, 2019. **37**(7).
27. Koçak, B., A.I. Fernandez, and H. Paksoy, Review on sensible thermal energy storage for industrial solar applications and sustainability aspects. *Solar Energy*, 2020. **209**: p. 135–169.
28. Bhutto, A.W., *et al.*, Promoting sustainability of use of biomass as energy resource: Pakistan's perspective. *Environmental Science and Pollution Research*, 2019. **26**(29): p. 29606–29619.
29. Wubben, E.F.M. and G. Nuhoff-Isakhanyan, The wicked problem of promoting sustainability by means of enhanced biomass utilization. *International Food and Agribusiness Management Review*, 2013. **16**(special issue): p. 45–50.
30. Jastrzębski, K. and P. Kula, Emerging technology for a green, sustainable energy promising materials for hydrogen storage, from nanotubes to graphene: A review. *Materials*, 2021. **14**(10).
31. Bin Abu Sofian, A.D.A., *et al.*, Advances, synergy, and perspectives of machine learning and biobased polymers for energy, fuels, and biochemicals for a sustainable future. *Energy and Fuels*, 2023.
32. Bin Abu Sofian, A.D.A., *et al.*, Machine learning and the renewable energy revolution: Exploring solar and wind energy solutions for a sustainable future including innovations in energy storage. *Sustainable Development*, 2024.
33. Włodarczyk, B., *et al.*, Assessing the sustainable development and renewable energy sources relationship in EU countries. *Energies*, 2021. **14**(8).
34. Zafar, M.W., *et al.*, The linkages among natural resources, renewable energy consumption, and environmental quality: A path toward sustainable development. *Sustainable Development*, 2021. **29**(2): p. 353–362.
35. Yang, G., *et al.*, Time for a change: Rethinking the global renewable energy transition from the Sustainable Development Goals and the Paris Climate Agreement. *Innovation*, 2024. **5**(2).
36. Nowotny, J., *et al.*, Towards global sustainability: Education on environmentally clean energy technologies. *Renewable and Sustainable Energy Reviews*, 2018. **81**: p. 2541–2551.
37. Child, M., *et al.*, Sustainability guardrails for energy scenarios of the global energy transition. *Renewable and Sustainable Energy Reviews*, 2018. **91**: p. 321–334.

38. Child, M., *et al.*, Corrigendum to Sustainability guardrails for energy scenarios of the global energy transition [*Renew. Sustain. Rev.* (2018) 91 321–334], (S136403211830176X), (10.1016/j.rser.2018.03.079). *Renewable and Sustainable Energy Reviews*, 2021. **138**.

39. Donnelly, K.B., Storing the future of energy: Navigating energy storage policy to promote clean energy generation. *Environmental Progress and Sustainable Energy*, 2023. **42**(2).

40. Anastas, P., *et al.*, The power of the united nations sustainable development goals in sustainable chemistry and engineering research. *ACS Sustainable Chemistry and Engineering*, 2021. **9**(24): p. 8015–8017.

41. Buonocore, J.J., *et al.*, Correction: Metrics for the sustainable development goals: Renewable energy and transportation. *Palgrave Communications*, 2019. **5**(1). doi: 10.1057/s41599-019-0336-4).

42. Buonocore, J.J., *et al.*, Metrics for the sustainable development goals: Renewable energy and transportation. *Palgrave Communications*, 2019. **5**(1).

43. Joshi, P.B., N. Chaubal-Durve, and C. Mohan, Full blown green metrics, in *Green Chemistry Approaches to Environmental Sustainability: Status, Challenges and Prospective*. 2023, Elsevier. p. 109–129.

44. Romero, J.C. and P. Linares, Exergy as a global energy sustainability indicator: A review of the state of the art. *Renewable and Sustainable Energy Reviews*, 2014. **33**: p. 427–442.

45. del Río, P. and M. Burguillo, Assessing the impact of renewable energy deployment on local sustainability: Towards a theoretical framework. *Renewable and Sustainable Energy Reviews*, 2008. **12**(5): p. 1325–1344.

46. Zou, X., *et al.*, How to achieve green development? A study on spatiotemporal differentiation and influence factors of green development efficiency in China. *PLoS ONE*, 2024. **19**(1 January).

47. Mahmood, S., *et al.*, Sustainable infrastructure, energy projects, and economic growth: Mediating role of sustainable supply chain management. *Annals of Operations Research*, 2024.

48. Elliott, D., Renewables: A review of sustainable energy supply options, in *Renewables: A Review of Sustainable Energy Supply Options*. 2013: IOP Publishing Ltd. p. 1–167.

49. Kelly, M.J., Lessons from technology development for energy and sustainability. *MRS Energy and Sustainability*, 2016. **3**(1).

50. Tian, J. and X. Chen, Physical separation: Reuse pollutants and thermal energy from water. *Water (Switzerland)*, 2023. **15**(6).

Chapter 2

PESTEL analyses of GSE

The most important human endeavor is the striving for morality in our actions. Our inner balance and even our very existence depend on it. Only morality in our actions can give beauty and dignity to life.

Albert Einstein [1]

2.1 INTRODUCTION

Energy is integral to the sustainability of Earth. Major energy scientific, legal, environmental, and political drivers are forces, as described in the recently published book, "Sustainable Agricultural Chemistry in the 21st Century" [2]. These forces represent areas of concern for global and societal survival. They present challenges and barriers to environmental and business development. Using a PESTEL (Political, Economic, Social, Technological, Environmental, and Legal) analysis, it is possible to examine the forces and how they can be used to strengthen and guide the desire to choose green sustainable energy (GSE) [3]. These areas within a PESTEL analysis capture the forces that shape and determine the overall effort to achieve sustainability. These forces are unifying the overall movement toward sustainability (see Figure 2.1).

Energy is critical to the growth and development of society. However, its harvesting and use need to be guided in the 21st century by the principles of sustainability [4]. This is indispensable to the achievement of global sustainable goals [5]. Within the demands for sustainability, there will be different approaches and methods used to achieve the goals. Which approaches are used and how they are developed are vital. The field of GSE must incorporate them. Consequently, GSE must develop its resources, methods, and practices to meet the goals of sustainability.

Most importantly, the PESTEL analysis mentioned above will guide GSE. This will be combined with life cycle and multi-criteria decision analyses to argue for their adoption. As this occurs, the efforts will become stronger and mature into circularity (see Chapter 3, "Green sustainable

DOI: 10.1201/9781003407447-4

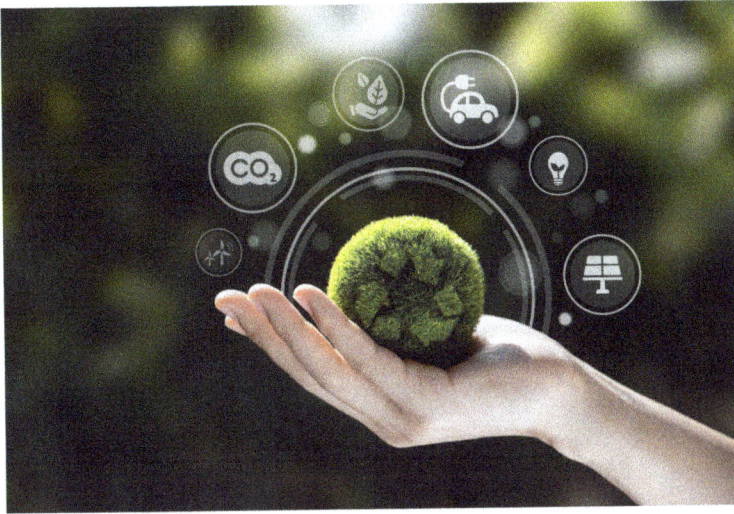

Figure 2.1 Human use of energy affects environmental sustainability.
Source: Shutterstock Photo ID: 2162001981.

energy-circularity nexus"). The importance of these ideas and their meaning for the work in GSE will become clear in this chapter.

2.2 IMPORTANT FORCES AFFECTING GSE

The global green sustainable energy effort is intentionally evolving. The benefits from the use of green energies are clear: reduce greenhouse gas emissions, diversify the energy supply, and lower dependence on volatile and uncertain fossil fuel markets. The future looks promising for green energy sources, and governments are trying to identify viable solutions to the energy crisis and reduce dependence on fossil fuels. Worldwide, there is a growing necessity for and support for developing green energy sources, and this will help accelerate the current energy transition (see Figure 2.2). Despite these positive developments, much remains to be done globally to make the energy transition a reality [6].

2.2.1 Four dimensions of strategic importance for sustainability

There are four main dimensions that are of strategic importance to the sustainable development of energy: energy resources, energy management, energy technologies/science, and energy use [7]. These areas must work in harmony to achieve sustainability. GSE needs to be guided by the six broad areas forming the PESTEL analysis. Based on the prevailing energy

Figure 2.2 Forces in GSE.

Source: Shutterstock Vector ID: 2297429939.

situation, a PESTEL analysis provides a means for addressing the political (P), economic (E), social (S), technological (T), environmental (E), and legal (L), challenges that constrain the development of GSE technologies [8].

Sustainability is critical to the harvesting, development, management, and use of energy. Historical energy use has achieved societal growth, but it has come with a high cost: pollution of our water, soil, and air, all of which pose human health and ecological threats [7]. The demand for sustainability provides challenges and opportunities. This requirement for sustainability must be guided by the development of useful policies, education, products, technologies, management procedures, and ethical principles that protect human health, well-being, and the environment, and also protect future generations. Realizing the finiteness and dangers of fossil energy systems, we have entered the age requiring adaptation and adoption of new energy systems.

2.2.2 Energy within the world system

If the sustainability of our planet is to be achieved, then our world must be treated as an integrated system. The health of the entire system is ultimately a response to and function of nature and human interactions with nature. This is no more strongly illustrated than by looking at energy (harvesting and usage). Research shows that the consequences of "energy abuse" are becoming more and more common in various geographical locations, and that will impact all sustainability efforts [9]. The benefits of our healthy use and the consequences of our abuses clearly show that energy is a constitutive element of our world's environment.

In this awareness of global interconnectedness and heightened environmental concern, definitive approaches are needed for all issues within the system. This is especially true for energy. It is necessary to anticipate and research new sources of energy to determine if they are green and sustainable. Both basic research and development of novel process concepts are urgently needed to enable the designing and implementation of environmentally safe, sustainable system solutions. The same thinking must be applied to harvesting, storage, distribution, and use of energy.

2.2.3 Identification of issues

All the challenges facing global sustainability are intertwined with energy (use of materials, cleaning of raw materials and products, heat removal, etc.). The critical issues center on the awareness of the finite nature of many resources, as well as the limited environmental tolerance toward human activities. The linear route of production, in which scarce resources are consumed and the concomitant production of waste, is the major contributing factor to global crises such as climate change, diminished biodiversity, as well as food, water, and energy shortages [10]. The production, use, and storage of energy play a role here, in the problem as well as its solution.

The dual problems of ecological crisis and global warming highlight the necessity of sustainability. Fossil fuel exhaustion and environmental pollution require an integrated plan to address the issues related to energy and the environment. Increasing population and growth in technology are creating higher demands for energy. Concerns are increasingly associated with the environmental impacts, life cycles, and supply chain sustainability of all sources of energy. Despite reports of many promising results, critical issues remain to be analyzed for any resulting cross-disciplinary problems. There is still a need for knowledge regarding the chemistry of processing conditions and precursor components, mechanisms of materials for energy harvesting and storage, and their corresponding electrochemical profiles [11].

2.2.4 A moral imperative for GSE

In his encyclical "Laudato Sì. On Care For Our Common Home," Pope Francis united the concern for our natural home (the earth) with a necessary obligation we should feel toward all aspects of nature [12]. This Encyclical Letter, issued in May 2015, contains some legal and economic aspects that go beyond purely religious relevance, touching upon the political, social, and ethical spheres. This should be a motivation to use knowledge to expand the understanding of ecology to include the physical earth and all its constituents (environment, society, economy, technology) [13]. Moral principles should be directing us to preferentially choose the actions and activities that preserve and sustain all dimensions of our global ecology. This supports a choice to pursue GSE. Three aspects of sustainable development (need, justice, and environmental limitation)are given equal importance [14].

If there is a universe of energy, how can we cleanly, effectively, and sustainably draw from it? The popular opinion that oil, gas, and coal are harmful, whereas renewables are clean, obscures the true sustainability challenge: material circulation. As we discuss later in circularity (Chapter 3, "Green sustainable energy-circularity nexus"), the ultimate challenge becomes producing and using energy in such a way that it is never exhausted, and no waste is produced. Environmental assessments, typified by the life cycle assessment (LCA), which assess the impact on the environment of the entire life cycle of a process, should provide initial information on the environmental impact of an energy source from its production to its use and help to identify opportunities for innovation which can also pinpoint which sources are most appropriate for GSE [15].

Sustainability becomes a moral imperative that requires a set of indicators to be established. The United Nations Sustainable Development Goals (SDGs) offer a detailed dashboard of sustainability indicators. However, the path to their achievement is difficult and will require many hard decisions. Sustainability needs to be expanded, including both a technical and a moral dimension. Sustainable development must be the result of moral imperatives (social, economic, and environmental). Operationally, priorities and gaps must be addressed to achieve sustainability [16].

Development and use of sustainable green energy by governments or the private sector should be undertaken because of an innate duty, and not simply out of self-interest. In other words, such actions should not be taken only because they will reduce costs, increase revenues, create jobs, or increase GDPs. Rather, they should also be taken to achieve the SDGs because, as rational human beings, preserving the Earth's environment and protecting the welfare of society as a whole are morally the right and good things to do [17] (see Figure 2.3).

2.3 ROLE OF ENERGY IN SUSTAINABILITY

Energy is at the heart of many of the previously mentioned Sustainable Development Goals – from expanding access to electricity, to improving clean cooking fuels, from reducing wasteful energy subsidies to curbing deadly air pollution that each year prematurely kills millions around the world. One of these goals – commonly known as SDG 7 – aims to ensure access to affordable, reliable, sustainable, and modern energy for all by the

Figure 2.3 Solar energy and sustainable resources.

Source: Shutterstock Photo ID: 535718161.

end of the next decade [18]. The adoption of energy-specific Sustainable Development Goals was a milestone in moving the world toward a more sustainable and equitable system.

Pursuing sustainable development in the face of climate change and environmental degradation is driving a significant shift toward green sustainable energy. A dependable, affordable, and stable GSE source must meet almost any future energy need. Wind, solar, hydropower, and biomass energy have benefits and challenges in reducing greenhouse gas emissions, mitigating environmental harm, and fostering long-term sustainability. Several innovative technologies inspired by these energy sources, such as solar power windows, energy-efficient buildings, smart grids, and their latest contributions to reducing environmental hazards, will become the visible ways in which energy is sustainably harnessed. Such innovations could revolutionize the energy sector and drive sustainable development globally [19].

Uncertainties, such as political conflicts between energy-producing nations, and use of conventional sources undermine the sustainability of energy security in conventional power generation systems. Green sustainable energy can provide a means to generate sufficient power to mitigate the risks and decelerate climate change. Compared to conventional sources, GSE is less affected by uncertainties such as political decisions and can reduce dependency on foreign energy resources. Reduced construction costs and advancements in GSE technology have helped the widespread deployment of these systems, enabling them to effectively meet power demand in areas of need. It is critical to transition to GSE to achieve sustainable and clean power while reducing the adverse environmental impacts of using conventional energy sources [20].

2.3.1 United Nations Sustainability Goals and GSE

Laudato si' was published a little before the United Nations ratified the agenda for the Sustainable Development Goals (*SDGs*) [21]. The 2030 Agenda for Sustainable Development with its SDGs and the Paris Agreement [22] seek to improve the well-being of people and the planet and strengthen the global response to the threat of climate change. The three most common policy-related issues slowing progress in achieving the SDGs regarding energy are lack of integrated/cross-sectoral planning, narrow emphasis on energy justice in policies, and the need for more cost-effective strategies in pursuit of the Paris Agreement. Research on the progress of implementation, impacts, and critical lessons from current policy efforts to achieve these global agendas is needed [23].

Clean, affordable, and efficient energy sources are necessary for a sustainable world. Energy crises, demand–supply mismatches, energy inequality, and high dependence on nonrenewable energy sources are challenges to the attainment of clean energy goals for sustainable development. Necessary for the achievement of the energy goals are:

- Policy reforms and better funding;
- Technology innovation and inclusion;
- Economic growth, rapid promotion of renewable, and alternative fuels.

Future research in energy sustainability should focus on energy justice, policy reforms, energy poverty, poor affordability, off-grid transmissions, renewable energy sources, alternative fuels, reforms in the energy supply chain, and international cooperation for better implementation of projects and for attracting foreign capital and private funds [24].

2.3.1.1 SDG 7

The United Nations' (UN) plans for 2030 include a renewable, affordable, and eco-friendly energy future. The 2030 agenda included 17 different Sustainable Development Goals for countries worldwide. The 7th SDG: Affordable and Clean Energy involved five main challenges:

- Limiting the use of fossil fuels.
- Migrating toward diversified and renewable energy matrices.
- Decentralizing energy generation and distribution.
- Maximizing energy and energy storage efficiency.
- Minimizing energy generation costs of chemical processes.

These challenges can be met only if existing technologies are fully implemented with the necessary international and national policies. Among the key solutions identified in addressing the five main challenges of SDG 7 are a global climate agreement; increased use of non-fossil fuel energy sources; Global North assistance and investment; reformed global energy policies; smart grid technologies; and real-time optimization and automation technologies [25]. Sustainable Development Goal 7 also requires innovations to incorporate it into common practice. Access to affordable, reliable, sustainable, and modern energy is the focus of SDG 7. It is underpinned by three targets:

- Ensuring universal access to energy services (7.1).
- Increasing the share of renewables in the energy mix (7.2).
- Improving energy efficiency (7.3) [26].

Reducing the human impact on the environment will necessitate a shift from traditional energy to GSE. It is desirable that these GSE sources and related production be viewed from a supply chain management perspective, so the integration of supply chain management will provide access to GSE production [30]. Against the background of continuously rising carbon emissions, accelerated energy transformation in developed countries must spread to developing countries to draw attention to energy security. Promoting the diffusion of GSE supply-chain management will be a powerful tool to support energy transformation, energy conservation, and emission reduction in movement to sustainability [31].

The efforts to achieve Goal 7 may face obstacles. Climate change impact scores show the most improvement. When this is used as a potential metric, it becomes possible for policymakers to identify, prioritize, and target specific sustainability ways to achieve SDG 7 [27]. Science-based metrics could allow corporations and interested investors to meaningfully align their actions with the SDGs in locations around the world where they can make the greatest positive impact [28].

2.3.1.2 GSE

GSE promotes SDG 7 by providing energy sources that will ensure energy security while supporting a global transition toward low-carbon energy sources. SDG 7 targets economy-wide absolute and per capita limits in overall energy use. This should precede adjustments in technology so that behavior can shift from energy excess for some to energy sufficiency for all [29]. This relates to the topic of green growth, including the use of alternative energy sources. Future green growth can be achieved only with the use of environmentally friendly renewable energy sources [30].

Current global energy production is mostly dependent on these fossil sources. Depletion of fossil resources and changes in the global price make it a major concern for sustainable use in the future. It will become critical to utilize environmentally safe and sustainable energy resources. The recent increase in the use of natural sustainable energy like solar power points to both the desirability and feasibility of GSE. Effective use of solar energy depends on the proper knowledge of its use and techniques. As we will discuss in later chapters, these GSE resources will necessarily be coupled with different storage technologies to obtain green sustainable energy generation [31].

2.3.2 Energy systems

Industrial development with the growth, strengthening, stability, technical advancement, reliability, selection, and dynamic response of the power system is essential. Governments and companies invest billions of dollars in technologies to develop, harvest, and distribute energy. They also require optimal energy planning and technologies to store energy [32]. In order for GSE to be fully incorporated into the global energy system, there will need to be concurrent work done in:

- Electricity production.
- Power delivery.
- Electric distribution networks.
- Energy storage.
- Energy saving, new energy materials, and devices.
- Energy efficiency and nanotechnology.
- Energy policy and economics.

Much of this work is being done already, but much more investment is needed in global research into artificial intelligence (AI) and data-driven models. In terms of power supply, AI can help utilities provide customers with renewable and affordable electricity from complex sources in a secure manner, while at the same time providing these customers with the opportunity to use their own energy more efficiently.

2.3.3 Energy and the four spheres

The process to resolve geopolitical and geo-economic tensions while achieving sustainability must be accomplished by transforming current economies into inclusive and sustainable societies. In the process, policymakers must be wary of not forgetting institutional practices of conserving and preserving ecosystems and biospheres with proactive and proper thinking. Economies must incorporate the challenge of a sustainable planet. Legal systems must work harder in the 21st century to use proper and critical thinking driven by an ecological conscience to preserve, conserve, and protect the environment that sustains us. The technology that is being built and fashioned to drive businesses must adhere to stringent ecological standards. Accomplishing Sustainable Development Goals could mitigate the debilitating effects of a globally warmer planet [33].

There is a need to understand the connection between all four spheres (atmosphere, lithosphere, geosphere, and hydrosphere [2]) and energy when using GSE as part of the climate change mitigation strategy. This need arises from the fact that GSE resources are often site-specific and require access and use of sectors of the environment. The impacts on the environment are significant, and they must be made in conjunction with the authority of local communities. These are integrated development policies when deciding on GSE projects. Requirements and planning policy, environmental impacts, and public consent are important for future research and policies. GSE with renewable resources acts as a modern technological fix but provides only a partial solution for the climate and energy crisis. Hence, there is need for environmental balance when moving toward sustainable development [34].

2.3.4 Energy distribution systems

The ability to effectively and sustainably distribute GSE will involve energy management systems (EMS). EMS must optimize energy consumption and reduce carbon footprints within various industries. By using EMS to efficiently monitor and control energy usage, identify areas of inefficiency, and adopt renewable energy sources, it can contribute to greener practices [35]. Smart grids are emerging as the solution for efficiently meeting the increasing energy demand. They adjust themselves to optimally deliver energy at the lowest cost and highest quality possible. The grid successfully makes use of renewable energy resources, electric vehicles, and smart pricing techniques, while attempting to achieve energy efficiency. It also promotes a greener environment by striving to reduce greenhouse gas emissions [36, 37].

Improvements in understanding and managing energy systems will include appropriate actions against global warming. Research focusing on the energy supply, end-use process, and distribution will provide new ways to reduce the environmental effects of energy systems. Energy storage, solar, wind, nuclear energy, and even hydrogen energy production have all grown in popularity and will need to be managed by these smart grids. The cost-benefit and mitigation potential of GSE generation will also benefit from the grids [38].

As GSE becomes more prevalent, EMS are undergoing significant change. Many countries have ambitions to increase the share of renewable energy in their energy mix. This presents the challenge of incorporating an increasing amount of volatile energy supply and a higher number of energy providers on the distribution grid level. The smart grid could be a solution to this challenge. However, the implementation of smart grid technologies is rather slow. Barriers to the implementation of smart grid technologies include:

- cost and benefit,
- knowledge, and
- institutional mechanisms as barrier categories [39].

2.4 A PESTEL ANALYSIS OF SUSTAINABLE GREEN ENERGY

A PESTEL analysis uses Political, Economic, Social, Technological, Environmental, and Legal, dimensions to evaluate a system [40] (see Figure 2.4). PESTEL analysis was initially designed in 1967 as a business planning tool. This methodical analysis assesses six external factors concerning the system, examining opportunities and threats arising from these factors. This method of analysis can evaluate GSE through a socio-political, techno-economic, legal, and environmental analysis approach, and illuminate the interrelation between technological facets and sustainable deployment [41].

2.4.1 PESTEL analysis framework

The PESTEL analysis framework is a strategic tool to assess and analyze the macro-environmental factors that can impact operations and decision-making. By examining these six categories, we can gain a better understanding of the external forces and trends that may affect GSE and develop strategies to adapt or capitalize on them. The following is an overview of each component and how it is applied to GSE.

PESTEL ANALYSIS

The PESTEL analysis is a management tool used to identify how company may get affected by external factors. It examines Political, Economic, Social, Technological, Environmental, and Legal factors.

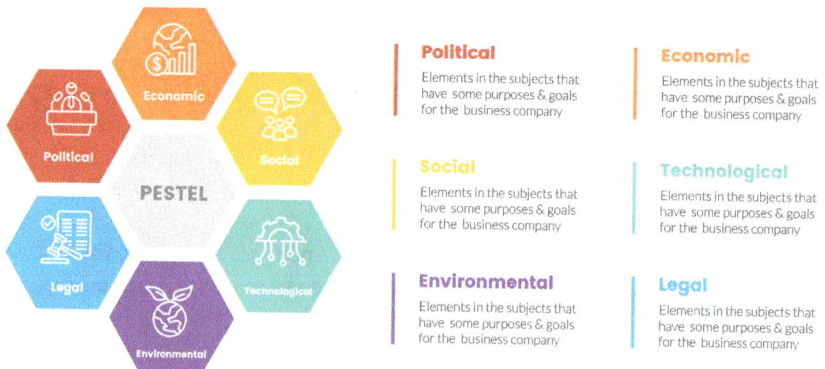

Political
Elements in the subjects that have some purposes & goals for the business company

Economic
Elements in the subjects that have some purposes & goals for the business company

Social
Elements in the subjects that have some purposes & goals for the business company

Technological
Elements in the subjects that have some purposes & goals for the business company

Environmental
Elements in the subjects that have some purposes & goals for the business company

Legal
Elements in the subjects that have some purposes & goals for the business company

Figure 2.4 PESTEL analysis strategy framework.
Source: Shutterstock Vector ID: 2283955891.

2.4.1.1 Political

This is one of the most important external factors affecting the adoption of GSE. The analysis of local, regional, and national political landscapes shows that these factors dominate if energy projects are started and/or maintained. Government policies directly affect the rate of business taxes, employee laws, general state of law and order, business compliances, and general ease of doing business. This will affect, for example, whether states or countries can economically support the harvesting and distribution of GSE.

2.4.1.2 Economic

GSE both depends upon and influences the economy of a geopolitical state. Economic influences include macro-financial factors such as state or national GDP growth, inflation or deflation rate, foreign currency debt, federal reserve interest rates and changes, and more. These factors will play a key role in determining target markets to sell to and where to distribute GSE. These economic factors may have varied effects depending on the type of energy and their ease of harvesting. For example, for raw wind power, a rise in the prices of the land or material will have an effect on the business. Often it could lead to a price hike of the final product, which is then borne by the consumer, which may reduce demand for the product/service.

2.4.1.3 Social

While less widely impactful than political or economic factors, social factors can potentially lead to significant effects on the success of GSE. For example, the acceptance and support of a solar farm would be very different in the Middle Eastern region than that of American or European regions. However, if the product or service is technical and aimed at improving the livelihood of the local populace, then it could be acceptable in multiple cultures. GSE technologies will have to be aware of the impact of social movements, cultural shifts, and sensitivities. This is a key impact that factors into enterprise human resource planning, product planning, and marketing content.

2.4.1.4 Technological

Technical or technological factors have a significant impact on any global effort, and they must be addressed if the project is to succeed. Technology impacts product development, harvesting and delivery efficiency, waste handling and reduction, and communication management efficiency. An effort that can take advantage of breakthroughs in tech and its impact on all areas of the project will also be able to better navigate continuity and growth to beneficially leverage these changes rather than get surprised by them. For example, the use of hydrogen as a GSE will require continual scientific and technological development to efficiently utilize this energy source.

2.4.1.5 Environmental

This dimension recognizes the real and measurable impact of environmental factors on how an operation is conducted. Factors such as CO_2 and global warming, natural disasters, availability of water and natural resources, and human and wildlife migrations can have a significant impact, especially on efforts in general and GSE in particular. This is an especially important factor for GSE, as it is central to the reason for its implementation.

2.4.1.6 Legal

These are external factors emerging out of political factors but are focused on regulations related to laws and environmental regulations, business conduct and operations, and taxation. An example would be property rights when there are questions regarding the use of natural resources for GSE.

2.4.2 Utility of a PESTEL analysis for GSE

A PESTEL analysis will enable a holistic understanding of the external forces shaping the adoption of GSE. Importantly, it can identify opportunities and threats through the analysis of factors, that impact the political, economic, social, technological, legal, and environmental conditions of the energy market.

Armed with insights from a PESTEL analysis, advocates of GSE can develop more informed strategic plans and make data-driven decisions. They can anticipate changes in the external environment and align their strategies accordingly. By considering external factors, proponents of GSE can better position themselves to capitalize on opportunities and minimize risks. By understanding the political, economic, social, technological, legal, and environmental landscape, it is possible to anticipate potential risks and develop strategies to mitigate them.

A PESTEL analysis helps to update information about changes in legislation, regulations, and compliance requirements. By staying ahead of regulatory developments, it is possible to adapt strategies and operations to ensure compliance and mitigate legal risks.

In summary, a PESTEL analysis empowers GSE to gain a comprehensive understanding of the external environment, identify opportunities and threats, make informed decisions, manage risks, and plan for long-term success. It's a vital tool for strategic management and navigating the complexities of the business landscape.

2.4.3 PESTEL and GSE

A proper sustainability analysis for GSE must include technical, economic, environmental, and social dimensions. In this regard, a look at the resource consumption in a particular energy process, like the green hydrogen process, is essential to guarantee that the process is indeed sustainable. To achieve the SDG by 2030 and net-zero CO_2 emissions by 2050, improving energy efficiency in all dimensions of energy is vital. A PESTEL analysis and Multi-Criteria Decision Analysis can identify the most advantageous process and the most promising resources. Such an analysis evaluates these systems in a comprehensive manner, including technology, CO_2 emissions, impacts on human health, ecosystems, and resources. Integrating green sustainable energy sources further achieves better results than fossil fuel consumption [42].

One of the specific objectives outlined in SDG 7 aims to ensure universal access to affordable, reliable, sustainable, and modern energy. The targets associated with this goal involve guaranteeing access to energy services that are affordable, reliable, and modern, as well as increasing the proportion of renewable energy sources, enhancing energy efficiency, and advancing technology for service delivery. Additional efforts are required to attain a renewable energy composition of 24.8% by the year 2030. It becomes very important to elucidate the ongoing energy transition in all regions of the world by examining aspects related to sustainability, considering PESTEL perspectives [43].

A PESTEL analysis helps in making strategic decisions for GSE. In each area of the analysis, there is a beneficial result. The PESTEL analysis carried out allows the definition of the requirements for the determinants of the sustainability goals under consideration [44]. This lays the foundation for a more balanced approach to the green sustainable energy effort. These strategies should reduce the impact of several processes on the environment through products, processes, and business policies using green applicable sustainable resources and environmental management systems. A holistic approach related to how to develop, implement, monitor, and improve a strategy (even an existing one) in the framework of sustainability designs is critical. This could be a useful tool for any policymakers, consultants, engineers, urban planners, academics, etc. The adoption of strategies will protect and enhance the Earth's natural capital and turn the environment into a resource-efficient, green, and competitive low-carbon economy in the near future [45].

2.5 LIFE CYCLE AND MULTI-CRITERIA DECISION ANALYSES

Sustainability is a concept that integrates at least three dimensions: environmental, economic, and social. Energy systems are key contributors to sustainability, so the methodology used for their evaluation must have indicators that are both quantitative and qualitative. It is therefore a challenge to choose the best methodology to accomplish this task. Life Cycle Analysis (LCA) and Multi-Criteria Decision Making (MCDM) combination are the right tools for the sustainability evaluation of GSE systems. They produce a set of sustainable indicators, evaluation methods, and the context where they are applied (such energy policies, electrical supply, and evaluation of projects). The hybrid framework of LCA and MCDM applied in combination appears as the most appropriate approach for this purpose [46].

2.5.1 Life cycle assessment (LCA) enables green transformation

Life Cycle Assessment (LCA) is a tool to support the development of green science. LCA is a standardized, structured, comprehensive, international environmental assessment tool and a rapidly evolving field of research that leverages and harmonizes efforts across many sectors to inform environmentally friendly solutions and choices. LCA is an evidence-based tool in the field of green science, being used as a decision-support tool at micro level (typically for questions related to specific products/processes) and macro levels (e.g., strategies, scenarios, and policy options) [47].

2.5.2 Multi-criteria decision-making

Energy management problems associated with rapid institutional, political, technical, ecological, social, and economic development have been of critical concern to both national and local governments worldwide for many decades. The application and use of MCDM approaches regarding energy management problems is invaluable, especially regarding the choice of GSE. The relevant topics span many different fields, including environmental impact assessment, waste management, sustainability assessment, renewable energy, energy sustainability, land management, green management topics, water resources management, climate change, strategic environmental assessment, and environmental management [48].

Environmental impact assessment is probably the most important area for applied decision-making. Decision-making approaches can help decision-makers and stakeholders in solving some problems under uncertain situations in environmental decision-making and these approaches have seen increasing application in various steps of the environmental decision-making process [48]. Green information technology and information system (ITIS) should be used to support these efforts that seek to prevent environmental pollution and attain sustainable development in collaborative enterprise (CE). By implementing green ITIS practices, improvements will result in green decisions on how to implement and adopt sustainable practices [49].

2.6 ECONOMY, SOCIETY, AND CULTURE

Green growth, globalization, and GSE consumption will impact environmental quality, and will effect economic growth [50]. There is growing attention from governments and regulators toward climate change and global warming, resulting in pressures to adopt the factors that make it possible for businesses to engage in green finance (GF). GF distributes financial resources to the research and development (R&D) of clean energy and environmentally friendly goods and processes; it is necessary for GSE adoption. GF policies help to alleviate the impacts of financial constraints involving new products, processes, services, and the global market [51].

Clean energy is becoming more affordable [52]. The successful transition of industrial operations from a conventional system to an environmentally friendly one is a prerequisite for achieving the Sustainable Development Goals discussed earlier. Nations must capitalize on novel technologies and eco-friendly policies to revive the natural climate. Green economic development should highlight the crucial role of GSE and environmentally friendly funding in a nation's sustainable development [53].

2.6.1 Circular thinking in GSE

Circularity can provide an effective way to accelerate the energy transition process by developing industrial collaborations and promoting synergies to enhance circularity and achieve economic, environmental, and social targets [54]. The global initiatives promoting energy transition expose the current supply chains to unprecedented energy-related issues. Being a green, independent, and secure alternative energy source, GSE enables organizations to develop resilience and fortify the circularity of the business ecosystem. Building awareness of the potential benefits of GSE, participating in energy-resilient practices, and promoting business models deploying environment-friendly technologies will foster sustainability and circularity in the supply chain operations [55].

2.6.2 Circular economy

A circular economy is "a model of production and consumption, which involves sharing, leasing, reusing, repairing, refurbishing and recycling existing materials and products as long as possible" [50]. It contrasts with the traditional linear economy. In a linear economy, natural resources are turned into products which are ultimately destined to become waste because of the way they have been designed and made. By contrast, a circular economy employs reuse, sharing, repair, refurbishment, remanufacturing, and recycling to create a closed-loop system, minimizing the use of resource inputs and the creation of waste, pollution, and carbon emissions [50].

The circular economy can play a key role in the transition to green sustainable energy. In fact, it can enable the shift to green sustainable power generation addressing the emerging security risks related to supply chains of minerals and metals crucial for GSE technologies. Circularity metrics are essential for assessing the effectiveness of actions taken and the progress toward measurable goals. Electricity generation systems, water circularity, the recycling rate of waste, and the usage of minerals and metals, as well as the need to enhance the installation of renewable capacity, are crucial areas for enhancing the circularity of the GSE system. This approach can support policymakers in planning strategies aimed at increasing the circularity of the electricity generation system at the national level [48].

2.6.3 Energy economy

Energy efficiency is crucial for achieving a balance between economic growth and sustainable development. As movement to carbon-neutral targets continues to gain momentum, green investments are becoming increasingly important. Therefore, it is critical to link green investment and energy efficiency. More advanced digital development is linked to improvements in energy efficiency. Furthermore, regions characterized by higher levels of digital economy exhibit a more pronounced contribution of green investment to energy efficiency [56].

2.6.3.1 Toward GSE

There are important interactions among financial, environmental, and economic issues and how they affect climate change. It is important to develop policies that support sustainable urbanization and increase economic resilience [57]. The demand for electricity is increasing, and most greenhouse gas emissions derive from the energy sector. Because of that, it is crucial to ensure the transition from fossil fuels to GSE. This energy contributes to sustainable development. Only the efficient implementation of circular economy and GSE practices can ensure that Sustainable Development Goals are achieved [58].

2.6.3.2 Developing the new paradigm

The traditional linear model of economics and GSE is problematic. To realistically achieve sustainability, a new circular model must replace it. Sustainable development of the economy and a harmonious relationship with the environment are priorities and provide guidelines for the world economy. In this context, it is important to define a holistic vision of the key principles of the development of sustainable economics and energy, which will inevitably involve green growth. The global economy must accept the transition to environmental friendliness, social responsibility, and good corporate governance. This should mean the transformation of the financial market to sustainable finance. Ecologically responsible banking activity as a format for sustainable development in the context of the transition in the world to the model of a green economy is a systematic analytical approach that will take into account the financial, economic, and environmental consequences of all dimensions of sustainable green energy [59].

Armed with the analyses described in this chapter, progress toward the adoption of GSE will accelerate.

REFERENCES

1. Einstein, A., *Albert Einstein: The Human Side.* 1981: Princeton University Press.
2. Nelson, W.M., Sustainable agricultural chemistry in the 21st century: Green chemistry nexus. *Sustainable Agricultural Chemistry in the 21st Century: Green Chemistry Nexus.* 2023: CRC Press. p. 1–294.
3. Sun, Z., *et al.*, Barriers to peer-to-peer energy trading networks: A multi-dimensional PESTLE analysis. *Sustainability (Switzerland),* 2024. **16**(4).
4. Ploeg, P., J. Revald Dorph, and N. Harvey, Planetary boundaries and sustainability principles: An integrated approach in the context of agriculture, Blekinge Institute of Technology Karlskrona, Sweden. 2016.
5. Minas, A.M., *et al.*, Advancing sustainable development goals through energy access: Lessons from the Global South. *Renewable and Sustainable Energy Reviews,* 2024. **199.**
6. Androniceanu, A. and O.M. Sabie, Overview of green energy as a real strategic option for sustainable development. *Energies,* 2022. **15**(22).
7. Duić, N., K. Urbaniec, and D. Huisingh, Components and structures of the pillars of sustainability. *Journal of Cleaner Production,* 2015. **88**: p. 1–12.
8. Zalengera, C., *et al.*, Overview of the Malawi energy situation and A PESTLE analysis for sustainable development of renewable energy. *Renewable and Sustainable Energy Reviews,* 2014. **38**: p. 335–347.
9. Belmonte-Ureña, L.J., *et al.*, Circular economy, degrowth and green growth as pathways for research on sustainable development goals: A global analysis and future agenda. *Ecological Economics,* 2021. **185.**
10. Keijer, T., V. Bakker, and J.C. Slootweg, Circular chemistry to enable a circular economy. *Nature Chemistry,* 2019. **11**(3): p. 190–195.
11. Feng, Y., *et al.*, Upgrading agricultural biomass for sustainable energy storage: Bioprocessing, electrochemistry, mechanism. *Energy Storage Materials,* 2020. **31**: p. 274–309.
12. Catholic, C. and K.W. Irwin, *On Care for Our Common Home: The Encyclical Letter laudato si'.* 2015, Paulist.
13. Montini, M. and F. Volpe, The need for an "integral ecology" in connection with the UN sustainable development goals, in *Care for the World: Laudato Si' and Catholic Social Thought in an Era of Climate Crisis.* 2019, Cambridge University Press. p. 56–67.
14. Mulia, P., A.K. Behura, and S. Kar, The moral imperatives of sustainable development: A kantian overview. *Problemy Ekorozwoju,* 2018. **13**(2): p. 77–82.

15. Jiménez-González, C., D.J.C. Constable, and C.S. Ponder, Evaluating the "Greenness" of chemical processes and products in the pharmaceutical industry: A green metrics primer. *Chemical Society Reviews*, 2012. **41**(4): p. 1485–1498.

16. Paoli, A.D., F. Addeo, and E. Mangone, Sustainability and Sustainable Development Goals (SDGs): From Moral Imperatives to Indicators and Indexes. A Methodology for Validating and Assessing SDGs, in *Perspectives for a New Social Theory of Sustainability*. 2020, Springer International Publishing. p. 47–68.

17. Salamat, M.R., Ethics of sustainable development: the moral imperative for the effective implementation of the 2030 Agenda for Sustainable Development. *Natural Resources Forum*, 2016. **40**(1–2): p. 3–5.

18. Matana Júnior, S., M. Antonio Leite Frandoloso, and V. Barbosa Brião, The role of HEIs to achieve SDG7 goals from Netzero campuses: Case studies and possibilities in Brazil. *International Journal of Sustainability in Higher Education*, 2023. **24**(2): p. 462–480.

19. Gayen, D., R. Chatterjee, and S. Roy, A review on environmental impacts of renewable energy for sustainable development. *International Journal of Environmental Science and Technology*, 2024. **21**(5): p. 5285–5310.

20. Kiehbadroudinezhad, M., et al., The role of energy security and resilience in the sustainability of green microgrids: Paving the way to sustainable and clean production. *Sustainable Energy Technologies and Assessments*, 2023. **60**.

21. Molina, M.C. and M. Pérez-Garrido, LAUDATO SI' and its influence on sustainable development five years later: A first LOOK at the academic productivity associated to this encyclical. *Environmental Development*, 2022. **43**.

22. Pekkarinen, V., Going beyond CO2: Strengthening action on global methane emissions under the UN climate regime. *Review of European, Comparative and International Environmental Law*, 2020. **29**(3): p. 464–478.

23. Akrofi, M.M., M. Okitasari, and R. Kandpal, Recent trends on the linkages between energy, SDGs and the Paris Agreement: A review of policy-based studies. *Discover Sustainability*, 2022. **3**(1).

24. Singh, S. and J. Ru, Accessibility, affordability, and efficiency of clean energy: a review and research agenda. *Environmental Science and Pollution Research*, 2022. **29**(13): p. 18333–18347.

25. Kay Lup, A.N., et al., Sustainable energy technologies for the Global South: challenges and solutions toward achieving SDG 7. *Environmental Science: Advances*, 2023. **2**(4): p. 570–585.

26. Liston, M., Powering our lives with secure, equitable and sustainable energy sources (SDG 7), in *Teaching the Sustainable Development Goals to Young Citizens (10–16 Years): A Focus on Teaching Hope, Respect, Empathy and Advocacy in Schools*. 2024, Taylor and Francis. p. 215–234.

27. Gebara, C.H. and A. Laurent, National SDG-7 performance assessment to support achieving sustainable energy for all within planetary limits. *Renewable and Sustainable Energy Reviews*, 2023. **173**.

28. Buonocore, J.J., et al., Metrics for the sustainable development goals: renewable energy and transportation. *Palgrave Communications*, 2019. **5**(1).

29. Burke, M.J. and R. Melgar, SDG 7 requires post-growth energy sufficiency. *Frontiers in Sustainability*, 2022. **3**.

30. Kayachev, G., *et al.* Expanding of Green and Renewable Energy as a Condition for Economy Transition to Sustainable Development. in *E3S Web of Conferences.* 2021. EDP Sciences.

31. Abdalla, A.N., *et al.*, Socio-economic impacts of solar energy technologies for sustainable green energy: A review. *Environment, Development and Sustainability,* 2023. 25(12): p. 13695–13732.

32. Ahmad, T., *et al.*, Energetics Systems and artificial intelligence: Applications of industry 4.0. *Energy Reports,* 2022. 8: p. 334–361.

33. Christie, L.G. and D. Cherian, Green economy and the future in a post-pandemic World, in *Integrated Green Energy Solutions.* 2023, Wiley. p. 1–10.

34. Lamhamedi, B.E.H. and W.T. de Vries, An exploration of the land–(renewable) energy nexus. *Land,* 2022. 11(6).

35. Kaur, J., Towards a sustainable Triad: Uniting energy management systems, smart cities, and green healthcare for a greener future, in *Emerging Materials, Technologies, and Solutions for Energy Harvesting.* 2024, IGI Global. p. 258–285.

36. Riaz, A., *et al.*, Review on comparison of different energy storage technologies used in micro-energy harvesting, wsns, low-cost microelectronic devices: Challenges and recommendations. *Sensors,* 2021. 21(15).

37. Anjana, K.R. and R.S. Shaji, A review on the features and technologies for energy efficiency of smart grid. *International Journal of Energy Research,* 2018. 42(3): p. 936–952.

38. Kang, J.N., *et al.*, Energy systems for climate change mitigation: A systematic review. *Applied Energy,* 2020. 263.

39. Muench, S., S. Thuss, and E. Guenther, What hampers energy system transformations? The case of smart grids. *Energy Policy,* 2014. 73: p. 80–92.

40. Thompson, S., Strategic analysis of the renewable electricity transition: power to the world without carbon emissions? *Energies,* 2023. 16(17).

41. Achinas, S., *et al.*, A PESTLE analysis of biofuels energy industry in Europe. *Sustainability (Switzerland),* 2019. 11(21).

42. Do Thi, H.T. and A.J. Toth, Environmental evaluation and comparison of hybrid separation methods based on distillation and pervaporation for dehydration of binary alcohol mixtures with life cycle, PESTLE, and multi-criteria decision analyses. *Separation and Purification Technology,* 2024. 348.

43. Revina Yasin, T., Z. Anna, and G. Lara Utama. Unpacking Indonesia's energy transition through a PESTEL analysis, for achieving sustainable development goals. in *E3S Web of Conferences.* 2024. EDP Sciences.

44. Sukiennik, M. and B. Kowal, Analysis and verification of space for new businesses in raw material market: A case study of Poland. *Energies,* 2022. 15(9).

45. Zorpas, A.A., Strategy development in the framework of waste management. *Science of the Total Environment,* 2020. 716.

46. Campos-Guzmán, V., *et al.*, Life cycle analysis with multi-criteria decision making: A review of approaches for the sustainability evaluation of renewable energy technologies. *Renewable and Sustainable Energy Reviews,* 2019. 104: p. 343–366.

47. Elnaggar, A., Nine principles of green heritage science: Life cycle assessment as a tool enabling green transformation. *Heritage Science,* 2024. 12(1).

48. Mardani, A., *et al.*, A review of multi-criteria decision-making applications to solve energy management problems: Two decades from 1995 to 2015. *Renewable and Sustainable Energy Reviews,* 2017. 71: p. 216–256.

49. Anthony, B., M.A. Majid, and A. Romli, Green information technology system practice for sustainable collaborative enterprise: A structural literature review. *International Journal of Sustainable Society*, 2017. **9**(3): p. 242–272.

50. Tran, H.V., Asymmetric role of economic growth, globalization, green growth, and renewable energy in achieving environmental sustainability. *Emerging Science Journal*, 2024. **8**(2): p. 449–462.

51. Agrawal, R., et al., Adoption of green finance and green innovation for achieving circularity: An exploratory review and future directions. *Geoscience Frontiers*, 2024. **15**(4).

52. Ma, X. and H. Najam, Achieving environmental sustainability goals through capitalizing on renewable energy channels: Role of green finance, resources productivity and geopolitical situation in the MENA region. *Geological Journal*, 2024.

53. Qing, L., et al., Investment in renewable energy and green financing and their role in achieving carbon-neutrality and economic sustainability: Insights from Asian region. *Renewable Energy*, 2024. **221**.

54. De Giovanni, P. and P. Folgiero, Strategies for the circular economy: Circular districts and networks, in *Strategies for the Circular Economy: Circular Districts and Networks*. 2023: Taylor and Francis. p. 1–144.

55. Mishra, R., et al., Renewable energy technology adoption in building a sustainable circular supply chain and managing renewable energy-related risk. *Annals of Operations Research*, 2023.

56. Dong, K., et al., Assessing the role of green investment in energy efficiency: Does digital economy matter? *Energy Exploration and Exploitation*, 2024. **42**(4): p. 1450–1471.

57. Niu, Y., Toward a greener energy transition: examining the effects of circular economy and carbon footprint for sustainable development. *Economic Change and Restructuring*, 2024. **57**(2).

58. Jakubelskas, U. and V. Skvarciany, Circular economy practices as a tool for sustainable development in the context of renewable energy: What are the opportunities for the EU? *Oeconomia Copernicana*, 2023. **14**(3): p. 833–859.

59. Lutsiv, B., et al., The role of green banking in ensuring the goals of sustainable economic development. *Financial and Credit Activity: Problems of Theory and Practice*, 2024. **1**(54): p. 23–36.

Chapter 3

Green sustainable energy-circularity nexus

We do not inherit the Earth from our ancestors, we borrow it from our children.

– Chief Seattle, Native American [1]

3.1 INTRODUCTION

A universe of energy must include the paradigm of circularity. Moreover, circularity requires green sustainable energy (GSE) to achieve sustainability (see Figure 3.1). Our appreciation of mass/energy has evolved considerably over the centuries, leading to the inevitable conclusion that the total amount of energy in the universe is constant. As was explained in the Introduction ("A Universe of Energy"), our challenge is to sustainably draw upon this energy source. It is recognized that the source of gravity in general relativity is ultimately the total energy in the universe of energy [2]. Within this system, energy maintains a balance with the total mass of the Universe [3].

Figure 3.1 Energy circularity.

Source: Shutterstock Photo ID: 1994233649.

DOI: 10.1201/9781003407447-5

The development, production, and use of GSE will allow us to satisfy our energy requirement in a holistic manner. The nexus of circularity and GSE forms an essential component in the accomplishment of sustainability.

Energy is a key component in a circular economy. The circular economy utilizes activities that depend upon sustainable energy and materials. Products suitable for long-term circulation, a primary characteristic in circularity, first need to be designed and manufactured which requires energy input. Even activities that are circular, such as remanufacturing and recycling, typically require large amounts of energy. No product can claim to be circular if it is manufactured or recirculated by consuming large amounts of finite and nonrenewable energy sources. Hence, it is crucial to prioritize energy efficiency and circular energy sources, such as GSE.

Circularity should also be viewed as a novel and innovative approach to synchronizing rapid economic growth, energy development/use, and raw materials preservation. The circular economy paradigm provides a sustainable foundation to enhance economic growth by avoiding waste, preserving natural capital, managing resource scarcity, recycling materials, maximizing energy efficiency, and recirculating them into the economy [4]. The circular nature of successful GSE, which will come from many potential sources, must also be judged by their socioeconomic and environmental impacts, including greenhouse gas emissions, land use, and community effects. To maximize energy production while curtailing environmental problems, ongoing research and recent developments are needed [5].

3.2 CIRCULARITY AND THE CIRCULAR ECONOMY

Circularity describes practices that optimize resource use and minimize waste across the entire production and consumption cycle, emphasizing sustainability and economic efficiency (see Figure 3.2). Circularity (CIR)

Figure 3.2 Circular energy.

Source: Shutterstock Vector ID: 2152122587.

presents an alternative to a linear model of economics. In a circular economy, resources can be used repeatedly, often for the same or similar purposes. The availability of GSE resources promotes the use of circularity in its movement to sustainability. CIR leads to a circular economy (CE), which provides an effective methodology for sustainable development. Sustainable consumption and production underlie the effectiveness of CIR/CE in the area of energy and will enhance the achievement of sustainable development goals [4].

Three major principles govern CIR/CE:

- Preserve and enhance natural capital (the world's stock of natural assets) by controlling finite resources and balancing the flow of renewable resources.
- Optimize resource yields by circulating products, components, and materials in use at the highest possible levels always.
- Make the system more effective by eliminating unintended negative consequences, like air and water pollution.

3.2.1 Circularity principles

Circularity principles resonate in the modern era. A circular economy showcasing green sustainable energy can reduce emissions and pressure on natural resources, chart innovative pathways to net-zero economies, create sustainable economic growth and jobs, and reduce supply chain risk GSE is harnessed to reduce costs and carbon emissions, further contributing to sustainability [6]. Political interest in the CE concept is increasing globally and creating the conditions for a transformation toward a more circular society. There is a spectrum of circular solutions that can address the challenges of food, material, energy, and water resource security. CIR/CE provides a framework that can be used to ensure an inclusive circular transformation that includes geopolitical, intragenerational, social, and environmental dimensions of circular solutions [7].

To support progress toward the transition to a CIR/CE, the ability to measure circularity is essential. The first step in coming to a suitable framework for GSE products is to define CIR/CE principles. There are six circularity principles that can be applied to GSE products:

- reduce reliance on fossil resources,
- use resources efficiently,
- valorize waste and residues,
- regenerate,
- recirculate, and
- extend the high-quality use of GSE.

To evaluate the circularity performance of GSE products with respect to these principles, what needs to be measured is defined considering both intrinsic circularity and the impact of this circularity. The intrinsic indicators provide a measure of success in the implementation of these circularity principles, and the subsequent impacts of circularity, i.e., impact of closing the loops on the accumulation of hazardous substances and the impact of circularity on sustainability (environmental, economic, and social). Yet, to unlock the potential of a sustainable circular bioeconomy, strong metrics are required [8].

3.2.2 Circularity and sustainability

Circularity is a sustainable model, process, or economic system focused on re-use and waste elimination. It's a method of achieving sustainability, and a representation of a sustainable ecosystem. A circular product or supply chain is likely to be more sustainable than a linear, non-circular one – but circularity alone isn't the only way to define something as sustainable.

Sustainability is an effective technique to reduce the effect of global warming on the earth. Looking at the concept of a green environment, various practices focused on environmental conservation and improved ecological health are ways to achieve the goal. These practices also include informed consumption, conservation measures, and investments in renewable energies. These are the basis of circularity. It is clearly important to integrate CE principles and sustainability metrics. To aid in this process, it will be important to (1) identify the circularity metrics used and their significance, (2) develop CE definitions anchored in the sustainability concept, and (3) provide recommendations on the basis of qualitative and quantitative results. This will enhance resource use efficiency and environmental stewardship. Moreover, sustainability metrics provide practical tools to assess and improve CIR/CE performance [9].

The use of GSE promotes sustainability across the supply chain and aids in the creation of a carbon-neutral bioeconomy. To improve GSE production, feedstocks and sources need to be identified and optimized for maximal production. In addition, the discovery of sustainable and low-cost catalysts certainly fosters the greening of the harvesting process, making it more cost-effective. This growth provides critical paths to achieve sustainability and circularity in the global energy market [10].

Sustainable Development Goals (SDGs) have been pivotal in identifying routes to carbon neutrality and sustainable solutions in environmental and urban development. There remains a critical need for innovative approaches to improving energy harvesting and reducing energy waste [14]. Adopting CIR/CE principles as a strategic pathway can mitigate environmental, social, and economic challenges and promote sustainable net-zero-energy solutions. There is a synergy between circular economies and advancement toward achieving the SDGs. Central to this are resource use, circularity in energy applications, and GSE integration.

The adoption of circular economy and life cycle thinking (LCT) tools also plays an important role in implementing and promoting sustainable development strategies. Four CE principles (reuse, recycle, reduction, and recovery) are effective when used as strategies to use innovative technologies, improve operational activities, and extend GSE. Environmental assessment, with some extension to economic and social impacts, helps to bring sustainability and circularity into alignment. Comprehensive, life cycle-based tools should be developed to thoroughly assess and improve circularity and GSE [15].

3.2.3 Circularity and energy

The connection between the energy sector and circularity can be referred to as the "energy-circularity nexus." On the one hand, the transition to a circular economy does not eliminate the need for energy. Rather, it prompts the efficient harvest/use energy, reduces primary energy consumption, and utilizes waste heat and renewable energy. On the other hand, the transition to GSE depends on the transition to a circular economy; currently, the rapid expansion of renewable energy infrastructure is increasing the demand for various critical minerals. Supply shortages are deemed likely in the next years. Therefore, the energy sector cannot afford to use scarce materials only once.

Circularity and the circular economy seek ways to reduce the environmental impact of energy systems by drawing from GSE sources, sustainable energy harvesting, reducing energy use, and minimizing waste generation. Designing and planning for sustainable/circular energy are essential considerations in the current era and must become integral to GSE [11]. The implementation of Sustainable and Circular Life Cycle Management should support efforts to eradicate and quantify waste, preserve the inherent value of products and materials, and encourage the adoption of GSE [12].

Driving this effort should be public policies toward sustainable and renewable energies that address the ultimate environmental challenges faced by the global environment. There are 4Rs (reduce, reuse, recycle, and recover) that are components of circular economic practices, and are often cited as the best route to sustainable development [13]. Commitments to the 4Rs of the CE must lead to the adoption and use of GSE, which will ultimately reduce greenhouse gas emissions. The circular economy and the utilization of GSE will have a positive economic return in terms of social well-being and the mitigation of environmental degradation.

3.2.4 Circularity promoting GSE

The goal of circularity and the circular economy is to transition from today's linear pattern of production and consumption to a circular system in which the societal value of products, materials, and resources is maximized over time. This goes beyond business and economics. CIR/CE do not ensure social, economic, and environmental performance (i.e., sustainability) on their own [14]. Green sustainable energy is a sustainable solution to the socioeconomic concerns associated with environmental issues and the depletion of non-renewable sources of energy. Resource circularity, increasing profits from green products, and designing processes for resource and energy efficiency have been found to be major sustainability criteria. Three main areas can allow the evaluation of CIR/CE promoting GSE. The first major area concerns the role of Environmental Management System (EMS) in increasing organizations' circularity; the second main area includes Ecodesign Directive and Environmental Technology Verification in the design and manufacturing processes of the products; and the third main area focuses on Green Public Procurement, Ecolabel, and Energy Label in driving greener consumption by setting products' circular criteria [15].

A transition to a circular economy will reduce pressure on natural resources and will create sustainable growth and jobs. This will lead to achieving climate neutrality targets and halting biodiversity loss. The European Commission adopted the new circular economy action plan (CEAP) in March 2020 [16]. It is one of the main building blocks for sustainable growth. The new action plan includes initiatives along the entire life cycle of products. It targets how products are designed, promotes circular economy processes, encourages sustainable consumption, and aims to ensure that waste is prevented and the resources used are kept in the EU economy for as long as possible.

The development of new technologies by manufacturing companies can also enable more circular solutions. Recently developed technology assessment tools enable (1) a quick screening of technical solutions or concepts, (2) the definition of decision-making criteria, and (3) the detection of potential negative environmental consequences of innovative technology, especially trade-offs related to circular economy. This provides a link to make informed decisions promoting CIR and CE improvements in GSE [17].

3.2.5 Energy supply chain in circularity

The global push to adopt GSE and energy transition exposes the current supply chains to unprecedented energy-related issues. Green sustainable energy technologies will enable organizations to build resilience and fortify the circularity of the business and environmental ecosystem. The direct effect of GSE on sustainable and circular supply chains will come in mitigating energy-related risks. The result will be that building awareness of the potential benefits of anticipating disruptions, participating in energy-resilient practices, and promoting business models deploying environment-friendly technologies will foster sustainability and circularity in the supply chain operations [18].

Transitioning the energy and industrial systems toward a circular economy will have key impacts on the availability and demand for alternative resources. Using a material flow analysis of industries reveals a close relationship among the material, energy, and emission flows of these sectors and their interdependencies. Analyses highlight the inherent interrelation between energy transition and circular economy. Transitioning to GSE has numerous techno-economic challenges and has substantial effects on production costs, but it also affects the availability of secondary materials. Circularity approaches can influence the energy transition by directly improving the availability of these materials. Future strategies to handle associated impacts and avoid material shortages will be an integral part of the harvesting/use of GSE [19].

3.2.6 GSE guiding circularity

Visionary action for integrating the adoption of GSE is crucial for developing CIR/CE. Accelerating the transition to GSE is supported by new advances in circular approaches to energy harvesting and use. Optimization of the energy-water and energy-air nexus increases the possible resources which become part of circularity. The same holds for green hydrogen and optimized hydrogen supply chains and components. Innovations in energy storage and applications span solar photovoltaic systems with hybrid energy storage, cooling for optimized and extended battery performance, and new energy storage designs for sector coupling. Upgrading and circularity in energy and materials based on hydrothermal carbonization of organic residues, upgrading processes for bio-aviation fuels, and advances in pyrolysis for fossil fuel replacement are other advances, alongside those in heat transfer and technologies. These advances provide critical insights in a time that calls for an important turning point for increasing global GSE capacity and the energy efficiency improvement rate [20].

Energy efficiency and appropriate, circular, energy sources are the key considerations when managing energy in the circular economy. Climate change, which is intrinsically linked to the need for energy both now and in the future, illustrates this connection. Sustainable green energy is a future solution, and many GSE technologies have been developed for different purposes. However, progress toward net zero carbon emissions by 2050 and the role of GSE in 2050 need to continue. GSE can help guide circularity considerations for world leaders and policymakers. Securing global energy leads to strong hopes for meeting the Sustainable Development Goals (SDGs), such as those for hunger, health, education, gender equality, climate change, and sustainable development. GSE can be a considerable contributor to future fuels [21].

The increased growth of sustainable green energy exploration and development will demand assistance from CIR/CE. The impact of GSE on circularity in the environment will be important. GSE effects on the environment will include economic growth, financial development, and energy. This would suggest that policies based on circular practices for energy generation can help achieve Sustainable Development Goals [22].

3.3 COMPONENTS OF CIRCULARITY-BASED GREEN SUSTAINABLE ENERGY

Developers and users of GSE have the potential to play a huge role in the strengthening of circularity and a circular economy. Many of the technological developments that could accelerate circularity are within their sphere of operations. These include innovations in materials composition and efficiency, electrification, hydrogen production, biochemistry and synthetic chemistry, and carbon capture and use.

As shown in Table 3.1, the components of circularity align with the tenets of GSE. As the global population and living standards rise, the demand increases for basic amenities including energy resources. Manufacturing automation has led to mass production and consumption, further increasing the demand for energy. The existing linear economy approach has led to increasing waste production and resource depletion, posing significant environmental and public health threats. To combat these limitations, the CIR/CE can protect the environment and improve economics by reducing energy and resource consumption. Thus, major impetus should be given to strengthening the backbone of the economy where tools such as green technologies, decarbonization strategies, material flow analysis, life cycle assessment, ecological footprints (water, carbon, and material), substance flow analysis, circularity indices, eco-designing, new business models, and policy play an essential role in the areas of socioeconomic sustainability, to enhance socioeconomic growth in a sustainable manner [23].

Table 3.1 Components of circularity in energy

Principles	Phases	Circularity dimensions
Prioritize renewable inputs	Production/ distribution	Strategies: • maximize the use of renewables • minimize value leakage across the value chain
Maximize product use	Consumption	Strategies: • sharing, • reusing, • repairing, • remanufacturing and • recycling
Recover by-products and waste	End of a product's life	The circular economy stops value leakage due to discarding products and materials after use.

3.3.1 Sustainable circular economy

The road to circularity is a path toward a more efficient and sustainable global society. Circular thinking decouples economic activity from the consumption of materials and energy by creating closed-loop cycles in which waste is minimized or even eliminated, and in which resources, including carbon, are reused. It does so by using resources efficiently, prioritizing renewable inputs, maximizing a product's or process' lifetime, and capturing and repurposing what was previously regarded as waste (see Table 3.1).

Sustainability is a strategic approach to support the environment and socioeconomic development. Circularity and the Circular Economy present an alternative paradigm to support market sustainability and deal with both environmental and socioeconomic challenges. These challenges almost demand that countries switch from linear economies to circular and sustainable economies [4].

A circular economy decouples growth from negative externalities by building on the following three key circular economy principles:

- Designing out waste and pollution
- Keeping products and materials in use
- Regenerating natural systems

Sustainability "evolved from the fields of ecology and environmental science...," and it describes all activities that ensure that human beings can co-exist with the natural world around them [24]. In his paper, "The Circular Economy – A new sustainability paradigm?", Martin Geissdoerfer suggests that both sustainability and circularity share a global perception, where all problems on a global scale must be addressed through shared responsibilities and by coordinating multiple agencies such as government, businesses, and individuals [25]. They both involve the people and the planet approach, but also the profit approach through innovation and design. Due to this reason, both concepts also show the potential to incur costs and risks but at the same time, they open up opportunities for value creation and diversification.

3.3.2 Role of energy in CIR and CE

Both the transition to GSE and the transition to a circular economy are fundamental to a sustainable future. A key principle in the circular economy is the cascading use of mass in products that create the most value over their lifetime. However, such "low-value" applications may be associated with greater environmental and socioeconomic benefits depending on the context [26]. Looking at organic waste-to-energy (OWtE) technologies, there is a steadily increasing driver in the establishment of an ever more efficient and sustainable circular economy. The advantages of OWtE processes are well known: not only do they reduce the waste volumes sent to landfills or incineration plants, but also and foremost, through the energy they yield (biogenic carbon dioxide, among others), they reduce dependence on fossil fuels [27].

CIR/CE emphasize activities that preserve value in the form of energy and materials. Energy will always be required in the circular economy. Many of the products available today may not be suitable for circularity because they are not long-lasting, repairable, recyclable, etc. Products suitable for long-term circulation first need to be designed and manufactured which will require energy input. Even activities that are circular, such as remanufacturing and recycling, typically require large amounts of energy. Most materials used in energy harvesting, such as structural foundations, will likely need to be disassembled, crushed, shredded, and/or melted to a secondary raw material before they can be used in new products at a later stage. Keeping materials in the loop during every single step in the process requires energy.

Lastly, no product can be labeled as circular if it is manufactured or recirculated by consuming large amounts of finite and nonrenewable energy sources. Hence, it is crucial to prioritize energy efficiency and circular energy sources, such as renewable energy or waste heat. A circular system is a key pillar of sustainability, security, and efficiency in the energy sector. The green sustainable energy transition requires incorporating the CIR/CE principles in the design process [4]. Importantly, citizen involvement in decision-making processes, valorization of suitable waste from an environmental perspective, and stability of political choices also play roles here [28].

3.4 GSE EXPANSION IN A CIRCULAR ECONOMY

There can be no circular economy without clean energy and no transition to GSE without a circular economy (see Figure 3.3). The two are intrinsically linked and depend on one another. This relationship is termed the GSE-circularity nexus. Shortages of rare and finite minerals would hurt the growth of GSE. While reducing the demand for fossil energy sources, the increase in GSE infrastructure produces a high demand for rare and finite minerals. For example, building solar photovoltaic (PV) plants or wind farms generally requires more minerals than establishing their fossil fuel-based counterparts.

Energy efficiency and appropriate, circular, energy sources are the key considerations when managing energy in the circular economy. A holistic approach to circularity in the energy sector is required to meet some of the sector's sustainability challenges. The focus will extend from energy efficiency and energy sources to aligning products and materials, technologies, and processes as well as strategies with circular principles:

Figure 3.3 LCA and circularity.

Source: Shutterstock Vector ID: 2279973729.

- **Energy efficiency:** Global energy demand is constantly growing and is expected to continue to grow. To reduce the primary energy demand, there is a continued need to gradually improve the energy efficiency of industrial processes with the aim of consuming as little energy as possible.
- **Green sustainable energy:** GSE should be the prioritized energy source for every use involving circularity or circular economy. This energy includes both energy that can be generated from natural forces, such as solar and wind as well as energy from biomass, such as agricultural or forestry residues.
- **Waste heat and industrial symbiosis:** Another way of producing "circular energy" is from industrial processes that produce waste heat. However, focusing on energy efficiency and energy sources alone is not sufficient to solve the sustainability challenges of our current energy system which include resource scarcity, waste, and carbon emissions.

The circular economy favors activities that preserve value in the form of energy and materials. Nevertheless, energy will always be required in the circular economy. Many of the products available today are not yet suitable for circulation because they are not long-lasting, repairable, recyclable, etc. Table 3.2 highlights circularity strategies for GSE.

3.4.1 Life cycle assessment tools help CIR/CE-energy expansion

The use of life cycle assessment (LCA) to evaluate CIR/CE-GSE nexus is very informative. Circularity/circular economy and sustainable development

Table 3.2 Circularity strategies for GSE

Area	Strategy	Examples
Circular products, parts, and materials	• Make products long-lasting, reusable, repairable, recyclable, etc. • Ensure that products actually re-enter the loop. (Product design for recycling)	Solar panels, wind turbines, batteries, etc. should live up to circular principles
Circular technologies and processes	• The energy sector develops and implements additional technologies and processes that allow for products, parts, and materials to re-enter the loop	Adequate reuse processes or recycling technologies at scale for wind turbine wings
Circular strategies	• Transform the energy sector toward the circular model • Systemic change from the choice of raw materials to design and manufacturing decisions • Maintenance programs and recirculation at end-of-life	Prioritize how to invest resources and develop targeted circular strategies

are not equivalent approaches and do involve some trade-offs. Circularity and sustainability are also complementary, although there are challenges related to the measurement of the sustainability performance of CIR/CE strategies. The contributions and limitations of life cycle assessment (LCA), including sustainability and circularity indicator-based approaches help to further understand how these different approaches and indicators could be appropriately deployed to develop more circular and sustainable product systems [29].

The adoption of CIR/CE and life cycle thinking (LCT) tools can play an important role in implementing and evaluating sustainable development strategies of GSE. The application of circular economy and LCT provides a means to analyze approaches and practices of applying circularity economy concepts such as circular economy principles, strategies, indicators, and business models for GSE. When the four circular economy principles are applied (reuse, recycle, reduction, and recovery), new strategies for energy emerge: using innovative technologies, improving operational activities, and extending the development of GSE [30]. Comprehensive LCTs should be developed for the full range of GSE identification, harvesting, and use to thoroughly assess and improve circularity and sustainability.

Circularity can be used to add value to all dimensions of GSE. The use of LCT has been used on biomass, but its application to the full range of GSE must increase. It can initially assess the harvesting and storage of energy from existing systems, considering mostly environmental concerns, by applying life cycle assessment and initially neglecting economic and social issues. Some challenges facing this effort are: the expansion of system boundaries, the consideration of more endpoints, the development and use of regional databases, the development of policies to encourage GSE, and the addition of economic and social issues [31].

CIR/CE are pivotal for GSE resource management, but added depth will come from including life cycle costing, life cycle assessment, life cycle cost–benefit, life cycle benefit analysis, and life cycle sustainability assessment. A model of sustainable and circular life cycle management based on specific performance indicators allows the environmental, social, and economic impact of GSE to be assessed throughout the life cycle of products and services [12].

3.4.2 PESTEL analysis of the CIR/CE-GSE nexus

A PESTEL (see Chapter 2, "PESTEL analyses of GSE") review of the CIR/CE-sustainability nexus clearly shows its relevance to GSE. Since energy will always be required in the circular economy, a PESTEL analysis assures GSE meets Sustainable Development Goal #7 [32]. GSE has not fully replaced fossil fuels to date. Without a major shift in the trillions of dollars of subsidies and investment away from fossil fuels to renewables, catastrophic climate change is predicted. Transitioning to a GSE economy will require an investment away from fossil fuels into all areas covered by the PESTEL spectrum. Green sustainable energy provides possibilities to realize sustainable development goals, climate stabilization, job creation, a green economy, and energy security with careful planning [33] (see Figure 3.4).

GSE power generation is expected to increase fivefold worldwide by 2040 [29]. The growth of this energy production requires assessments of its role in tackling climate change and other impacts within complex environmental, economic, and social systems. The role of a PESTEL analysis is clear

Figure 3.4 Moving toward circular energy.

Source: Shutterstock Vector ID: 2165942733.

here. Life cycle environmental impact indicators (global warming, acidification, eutrophication, human toxicity, ozone depletion, photochemical oxidation, and cumulative energy demand) and economic sustainability will be valuable via life cycle costing. Social dimensions should be evaluated in the areas of public acceptance, technology safety, and local employment rate, although none rigorously considered a life cycle approach. Finally, the circular economy can improve the effectiveness of future research with studies on filling major data gaps in literature such as the lack of detailed documentation for specific materials and background process choices in life cycle assessment databases.

Future work must be directed toward the harmonization of the LCA with the PESTEL and CIR/CE-sustainability nexus. This will drive the analysis of multiple impacts and the development of dynamic environmental analyses to consider the long-term uncertainty due to climate change, data availability, and energy decarbonization [30].

3.5 TRANSITION TO GSE CIRCULARITY AND THE CIRCULAR ECONOMY

CIR and CE are vital to the acceptance and growth of GSE. CIR/CE will decouple economic development from the linear dynamics of finite resource extraction, use, and disposal [34]. Through incorporation of GSE circular economy can promote sustainability and help to tackle climate change.

The world can maximize chances of avoiding dangerous climate change by moving to a GSE circular economy, allowing countries to meet the goals of the Paris Agreement on Climate Action [35]. The CIR/CE strategy can reduce material and human footprint by introducing systemic solutions. However, there is still a huge gap remaining, since almost 90% of the world resources which enter the production chain are wasted. To fill in this gap, it is necessary to tackle the barriers that prevent the world from implementing and improving circular initiatives.

Green sustainable energy harvesting and use are integral to reducing greenhouse gas emissions and mitigating climate change while achieving the CIR/CE-sustainability nexus. GSE is necessary and has immense environmental, technical, and economic potential as the industry matures and business cases are proven. The movement toward adoption of GSE is first to identify a few key technical, economic, and regulatory goals that must be met. Secondly, factors that impede current GSE management efforts within the circular economy gap and those that can support sustainable technology deployment need to be addressed. Finally, there must be global communication to profit from all efforts in the nexus [36].

GSE technologies and supporting infrastructures are resource intensive: large amounts of critical raw materials, composites, plastics, metals, and concrete are required in the construction of renewable energy production plants. Therefore, these energy technologies must be developed and managed in the most environmentally, economically, and socially sustainable way. To this end, circular economy strategies can narrow, slow, and close resource loops in helping to achieve these goals [37]. However, technology design for the circular economy demands a better understanding of resource requirements, material alternatives, use performance, and end-of-life solutions to identify the most suitable life cycle management strategies.

Other CIR/CE-GSE strategies to be explored will include:

- Material substitution.
- Permanent magnets substitution.
- Generators substitution.
- Material efficiency improvements.
- Reuse.
- Recycling.

3.5.1 The path to GSE

Circularity and the circular economy will expedite the shift to GSE power generation by addressing the emerging security risks related to supply chains of minerals and metals crucial for green energy technologies. Circularity metrics show the effectiveness of actions taken and the progress toward measurable goals. The path involves energy policies, materials, energy, waste flows, and environmental sustainability. CIR /CE thinking have become an effective pathway for sustainable development [4]. However, to increase circularity while improving global sustainability, investment in and adoption of GSE are necessary. Ultimately, to transition toward a sustainable circular nexus involving GSE, a global commitment to this effort is essential [26].

Circularity and the circular economy are tools to keep the value of natural resources. Therefore, accomplishing effective strategic plans that are already directed to circularity, ensuring stakeholders' involvement, and providing sufficient funding can benefit circular economy development [38]. However, the mainstream approach and technocratic tradition of research and policy vis-à-vis energy transitions could result in the perpetuation of social inequalities, energy injustices, and the passive participation of citizens. In order to successfully deploy GSE a radical approach will need to be adopted [39].

3.5.2 Energy becoming GSE

Social innovation, circularity, and energy transition may all be considered environmental, social, and governance (ESG) components from a sustainability perspective. Implementing ESG practices requires social innovation, circularity, and energy transition. In the words of the United Nations

Table 3.3 Five critical actions to speed adoption of GSE

Action	Result	Barrier
Make GSE technology a global public good	Essential technologies such as battery storage systems allow energy flow from renewables.	Roadblocks to knowledge sharing and technological transfer remain.
Improve global access to components and raw materials	A robust supply of renewable energy components and raw materials is possible.	Significant international coordination to expand and diversify is lacking.
Level the playing field for GSE technologies	Technology, capacity, and funds for renewable energy transition exist, reducing market risk and enabling investments.	Domestic policy frameworks must urgently be reformed for more renewable energy projects and investments.
Shift energy subsidies from fossil fuels to green sustainable energy	Cuts emissions, and contributes to sustainable economic growth, job creation, better public health, and more equality.	Fossil fuel subsidies are one of the biggest financial barriers hampering the world's shift to GSE.
Triple investments in renewables	The reduction of pollution and climate impact alone could save the world up to $4.2 trillion per year by 2030.	At least $4 trillion a year needs to be invested in GSE and renewable energy to allow us to reach net-zero emissions by 2050.

Secretary-General, "renewables are the only path to real energy security, stable power prices and sustainable employment opportunities" [40].

There are five critical actions the world needs to prioritize now to transform our energy systems and speed up the shift to green sustainable energy (see Table 3.3) [41].

3.5.3 CIR/CE-GSE

Circularity and the circular economy can be seen as one approach or tool to tackle the transition toward GSE. This means that the circular model must address the economic, social, and environmental impacts involved in the development, harvesting, and use of energy. GSE does not necessarily enable circularity, and a circular business model does not promise the improvement of social, environmental, and economic factors that accompany GSE. Together, however, they will ensure a steady movement to sustainability.

REFERENCES

1. Nehme, B., et al., Photovoltaic panels life span increase by control, in *Predictive Modelling for Energy Management and Power Systems Engineering*. 2020, Elsevier. p. 27–62.

2. Melia, F., Inertia, gravity and the meaning of mass. *Physica Scripta*, 2024. **99**(1).

3. Annila, A., The substance of gravity. *Physics Essays*, 2015. **28**(2): p. 208–218.

4. Ghazanfari, A., An analysis of circular economy literature at the macro level, with a particular focus on energy markets. *Energies*, 2023. **16**(4).

5. Ali, F., *et al.*, Fueling the future: Biomass applications for green and sustainable energy. *Discover Sustainability*, 2024. **5**(1).

6. Singh, S.R. and D. Singh. Impact of circular economy on sustainable inventory model with renewable energy under green environment, in *Lecture Notes in Networks and Systems*. 2024. Springer Science and Business Media Deutschland GmbH.

7. Petelin, E., Security priorities in circular economy: A conceptual review. *Sustainable Production and Consumption*, 2024. **47**: p. 655–669.

8. Vural Gursel, I., *et al.*, Defining circular economy principles for biobased products. *Sustainability (Switzerland)*, 2022. **14**(19).

9. Shaikh, M.B.N., *et al.*, Metrics for sustainability and circular economy practices in context to modern manufacturing environment. *Circular Economy and Sustainability*, 2024.

10. Sreeharsha, R.V., N. Dubey, and S.V. Mohan, Orienting biodiesel production towards sustainability and circularity by tailoring the feedstock and processes. *Journal of Cleaner Production*, 2023. **414**.

11. Rosen, M.A., The Circular Economy and Energy, in *CSR, Sustainability, Ethics and Governance*. 2022, Springer Nature. p. 133–149.

12. Basile, V., N. Petacca, and R. Vona, Measuring circularity in life cycle management: A literature review. *Global Journal of Flexible Systems Management*, 2024. **25**(3): p. 419–443.

13. Nunes, A.M.M., *et al.*, Public policies for renewable energy: A review of the perspectives for a circular economy. *Energies*, 2023. **16**(1).

14. Walzberg, J., *et al.*, Do we need a new sustainability assessment method for the circular economy? A critical literature review. *Frontiers in Sustainability*, 2020. **1**.

15. Marrucci, L., T. Daddi, and F. Iraldo, The integration of circular economy with sustainable consumption and production tools: Systematic review and future research agenda. *Journal of Cleaner Production*, 2019. **240**.

16. Ritter, M., *et al.*, Towards achieving the sustainable development goals: A collaborative action plan leveraging the circular economy potentials. *Gruppe. Interaktion. Organisation. Zeitschrift fur Angewandte Organisationspsychologie*, 2024. **55**(2): p. 175–187.

17. Parolin, G., *et al.*, Enabling environmental sustainability and circularity assessment in technology development: The value-impact scanner. *Sustainable Production and Consumption*, 2024. **49**: p. 92–103.

18. Mishra, R., *et al.*, Renewable energy technology adoption in building a sustainable circular supply chain and managing renewable energy-related risk. *Annals of Operations Research*, 2023.

19. Abdelshafy, A. and G. Walther, Exploring the effects of energy transition on the industrial value chains and alternative resources: A case study from the German federal state of North Rhine-Westphalia (NRW). *Resources, Conservation and Recycling*, 2022. **177**.

20. Krajačić, G., et al., Sustainable development of energy, water and environment systems in the critical decade for climate action. *Energy Conversion and Management*, 2023. **296**.
21. Finecomess, S.A. and G. Gebresenbet, Future green energy: A global analysis. *Energies*, 2024. **17**(12).
22. Usman, M., et al., Contribution of energy based circularity for better environmental quality: An evidence from Bias-corrected linear dynamic approach. *Discover Sustainability*, 2024. **5**(1).
23. Mukherjee, P.K., et al., Socio-economic sustainability with circular economy: An alternative approach. *Science of the Total Environment*, 2023. **904**.
24. Owusu, P.A. and S. Asumadu-Sarkodie, A review of renewable energy sources, sustainability issues and climate change mitigation. *Cogent Engineering*, 2016. **3**(1).
25. Geissdoerfer, M., et al., The circular economy: A new sustainability paradigm? *Journal of Cleaner Production*, 2017. **143**: p. 757–768.
26. Chitaka, T.Y. and C. Schenck, Developing country imperatives in the circular bioeconomy: A review of the South African case. *Environmental Development*, 2023. **45**.
27. Zueva, S., et al., Review of Organic Waste-to-Energy (OWtE) technologies as a part of a sustainable circular economy. *Energies*, 2024. **17**(15).
28. D'Adamo, I. and C. Sassanelli, A mini-review of biomethane valorization: Managerial and policy implications for a circular resource. *Waste Management and Research*, 2022. **40**(12): p. 1745–1756.
29. Saidani, M. and H. Kim, Nexus between life cycle assessment, circularity, and sustainability indicators—Part I: A review. *Circular Economy and Sustainability*, 2022. **2**(3): p. 1143–1156.
30. Longo, S., et al., Circular economy and life cycle thinking applied to the biomass supply chain: A review. *Renewable Energy*, 2024. **220**.
31. Ramos Huarachi, D.A., et al., Life cycle thinking for a circular bioeconomy: Current development, challenges, and future perspectives. *Sustainability (Switzerland)*, 2023. **15**(11).
32. Niyogi, S. and S. Dinda, Achieving targets of SDG 7 in post-COVID-19: Critical review of recent Indian energy policies, in *International Trade, Economic Crisis and the Sustainable Development Goals*. 2024, Emerald Group Publishing Ltd. p. 149–160.
33. Thompson, S., Strategic analysis of the renewable electricity transition: Power to the world without carbon emissions? *Energies*, 2023. **16**(17).
34. Velasco-Muñoz, J.F., et al., Circular economy implementation in the agricultural sector: Definition, strategies and indicators. *Resources, Conservation and Recycling*, 2021. **170**.
35. Fux, H., What is the ideal scenario for circular economy to occur? A case study of the circe project. *Brazilian Journal of Operations and Production Management*, 2019. **16**(1): p. 157–165.
36. Woo, S.M. and J. Whale, A mini-review of end-of-life management of wind turbines: Current practices and closing the circular economy gap. *Waste Management and Research*, 2022. **40**(12): p. 1730–1744.
37. Bocken, N.M.P., et al., Product design and business model strategies for a circular economy. *Journal of Industrial and Production Engineering*, 2016. **33**(5): p. 308–320.

38. Tleuken, A., *et al.*, Legislative, institutional, industrial and governmental involvement in circular economy in Central Asia: A systematic review. *Sustainability (Switzerland)*, 2022. **14**(13).
39. Nguyen, M.T. and S. Batel, A critical framework to develop human-centric positive energy districts: Towards justice, inclusion, and well-being. *Frontiers in Sustainable Cities*, 2021. **3**.
40. Mitchell Crow, J., Achieving UN climate goals needs purposeful, persistent action from science. *Nature*, 2023.
41. Saleh, H.M. and A.I. Hassan, The challenges of sustainable energy transition: A focus on renewable energy. *Applied Chemical Engineering*, 2024. **7**(2).

Chapter 4

GSE metrics for resilience

Measure what is measurable and make measurable what is not so.

Galileo Galilei [1, 2]

4.1 INTRODUCTION

The sustainability of energy depends on three major pillars: economic performance, environmental effects, and social impacts (see Figure 4.1). To design green sustainable energy (GSE) using renewable sources and/or to compare existing energies and technologies in view of their sustainability, including energy consumption in chemical reactions, clear metrics must be developed [3, 4]. To remain competitive in the 21st-century global marketplace for all consumers, GSE must be recognized as in harmony with standards that define energy and sustainability. To accomplish this, metrics are required for measuring progress toward sustainability objectives [4].

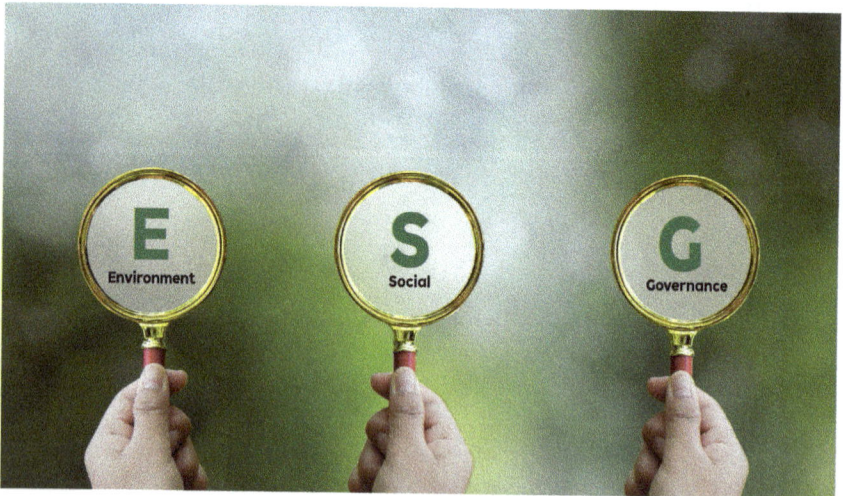

Figure 4.1 Measuring the value of green sustainable energy.

Source: Shutterstock Photo ID: 2322903721.

DOI: 10.1201/9781003407447-6

There is a connection between the second law of thermodynamics and environmental impact: when energy efficiency is increased, environmental effects are reduced. This trade-off relationship occurs because less energy is needed when it is used efficiently. Metrics on impact analysis of influential factors in energy are needed. Identification and standardization of metrics categories such as the materials used by the process and generated wastes, compounds' quality in terms of hazards and safety, energy consumption and economy, and the metrics used to elaborate a diagnosis and quantify the greenness will all be helpful. An understanding of the relationship between exergy, environment, and energy reveals the underlying, fundamental forces involved with improving efficiency [5].

The definition of GSE should lead to the metrics necessary to quantify them. This definition results from the basic ideas and principles of green chemistry and engineering. However, with the growing demand to continue to maintain energy for an increasing population, new sources of energy must be sufficient while being sustainable. This, in turn, leads to the need to measure the impacts at all levels to be able to manage and adapt new approaches which should be evolutionary in nature and not just be concerned with the yield and financial aspects. Quantification of the "sustainable-greenness" of GSE will require several parameters so as to not miss any of the target principles [6]. In this chapter we will examine the importance of metrics, the desired areas of energy and GSE to measure, and the available/possible metrics that will be most beneficial.

4.2 ENERGY IN OUR WORLD

There are different sources and types of energy (Figure 4.2). There are primary energies, those that are obtained directly from nature before being transformed, such as solar, wind, hydraulic, geothermal, or sea energy, as well as those contained in biomass, oil, natural gas, or coal. There are also secondary energies, which are obtained from the transformation of primary energies. This group would include gasoline, electricity, gas oil, fuel oil, etc. A third class is harvested energy, which is discussed in Chapter 8, "GSE harnessing, harvesting, and storage." There are also different ways of classifying types of energy based on their availability: either renewable, those that come from natural resources and are almost inexhaustible, or nonrenewable, those that have limited reserves that diminish with consumption.

Sustainable energy includes any energy source that cannot be depleted and can remain viable forever. They do not need to be renewed or replenished; sustainable energy meets the demand for energy without any risk of exhaustion. Furthermore, sustainable energy doesn't harm the environment (or at most, there is a minimal risk), increase climate change, or is too costly. On the other hand, renewable energy is theoretically exhaustible – it

Figure 4.2 Energy for the world.

Source: Shutterstock Photo ID: 121306519.

uses resources from the earth that can naturally be replenished, such as crops and biomatter. A renewable energy source like bioenergy uses biological masses (e.g., agricultural by-products like straw and manure) to create energy.

Figure 4.3 Indicators of sustainability.

Source: Shutterstock Vector ID: 2217237563.

4.2.1 Strongest sustainability indicators

A key driver of sustainability is the transition to sustainable energy resources. The sustainability impact assessment of energy production needs to evaluate all potential impacts on society, the environment, and the economy, as well as find a reasonable solution to shift toward GSE. Given the multitude of quantitative and qualitative methodologies for sustainability analysis, alongside the absence of a unified procedure, selecting an appropriate method poses a significant challenge in conducting this task. While there is a deficiency in a standardized approach for sustainability evaluation within electrical technology, it is possible to do an impact assessment using various parameters and dynamic methods with multiple temporal accuracies in a static life cycle. Using this methodology, the results indicate that hydropower, gas, and solar technologies exhibit the highest sustainability scores, respectively [7].

In the context of a rapidly growing energy demand and concerns about global climate change, GSE is a long-term solution toward secured energy supply and for the reduction of greenhouse gas emissions. However, climate change will certainly not be the only environmental issue. Should competitiveness between applications be considered in a sustainability assessment? In addition to responsible resource management indicators, many other aspects must be taken into account in order to achieve a complete sustainability assessment, especially recyclability, viability of the energy industry, equilibrium along the value chain, or social indicators such as social acceptability. Designing sustainability metrics is a new and complex research field. The whole value chain has to be evaluated and all dimensions (environmental, economic, and social) need to be explored [8].

The use of sustainability to measure energy has merit when addressing the mounting global challenges of environmental degradation and resource depletion. Indicators of sustainability focusing on energy are crucial tools used to assess and monitor progress toward achieving a more sustainable energy system. These indicators provide valuable insights into the environmental, social, and economic dimensions of energy practices and their long-term impacts. By analyzing and understanding these indicators, policymakers, businesses, and communities can make informed decisions, formulate effective policies, and steer their efforts toward a more sustainable energy future. These indicators serve as navigational guides, steering the world toward energy practices that support both present needs and the well-being of future generations [9].

4.2.2 21st-century sustainability

The increasing global importance of climate change forcefully drives the need for 21st-century sustainability. The identification of current practices and future needs will highlight desired characteristics of energy. Life cycle design optimization with consideration of sources, harvesting, and use is important for the sustainable development of resource-based circular economy. Finally, future design should be adopted to generate social-technical solutions that encompass both environmental sustainability and human well-being [10].

Evaluating what types of energy are sustainable in the 21st century is not a simple task. The entire life cycle of each energy source must be considered before it can be classified as sustainable. Within each life cycle there exist multiple indicators that should be included. Each of these indicators may impact or modify some phase of an energy source's life cycle that affects whether it is truly sustainable. Since the most recognized examples of sustainable technologies using renewable resources are wind power, biomass, and solar, these can provide examples. Tools have been developed to assess the resiliency of natural environments to disturbances. However, most of these tools have not been developed to measure social sustainability in the context of the natural environment. This makes it extremely difficult in the face of climate change to include metrics for evaluating the impact of our energy choices on our society and societal vulnerability. But if social and cultural metrics are included in energy evaluations, energy choices can then be linked to not only environmental impacts but also social impacts. Sustainable energy consumption does not equate to sustainable societies, but the inclusion of both cultural values and ecological health in sustainability will ensure the most complete measure of sustainability [11].

4.2.3 Resiliency – defining energy characteristics in the 21st century

Resiliency is the capacity to recover quickly from difficulties, while sustainability is the ability of a system to be maintained at a certain rate or level. Energy resilience is the ability of the entire system to withstand and rapidly recover from energy shortages or power outages and continue operating with electricity, heating, cooling, ventilation, and other energy-dependent services. GSE can help prevent electric grid disturbances and enable fast recovery after a disturbance. Using green sustainable energy resources – solar, water, wind, geothermal, and bioenergy – and enhanced power electronics provides more ways to maintain the energy supply or facilitate recovery after an outage.

Energy is vital to security, emergency services, critical infrastructures, and the economy. Resilience of the energy system against high-impact low-probability events is of particular importance to ensure the stability and reliability of the system planning and operation. Short- and long-term plans with different categorizations are critical for long-term resiliency. Short-term plans refer to resilience-oriented scheduling, and long-term plans indicate fundamental corrections such as hardening and equipment upgrades [12]. An integrated energy system characterized by multi-energy complementary and supply-demand interaction can promote the resilience of urban energy systems. Incorporating the impact of extreme events into the early-stage energy system planning is crucial to improve its resilience. While considering the uncertainty of GSE and demand response, an optimal planning method can deal with both economic performance and resilience of an integrated energy system under extreme natural disasters. The marginal cost of resilience improvement increases exponentially with the advancement of system resilience [13].

Access to affordable, reliable, and clean energy is crucial for fostering economic, social, and sustainable development. However, conventional power systems are grappling with numerous challenges, with climate change standing out as a major concern. Energy identification, development, and users must enhance their existing energy infrastructure and embrace alternative energy sources with lower carbon footprints. In this area, microgrids are playing a pivotal role in both adaptation and mitigation strategies. Smart grids are electricity networks that use digital technologies, sensors, and software to better match the supply and demand of electricity in real time while minimizing costs and maintaining the stability and reliability of the grid. A microgrid, on the other hand, is a self-sufficient energy system that serves a discrete geographic footprint, such as a college campus, hospital complex, business center, or neighborhood. Microgrids have a positive impact on the energy system in the environment and society. Microgrids need further development before fully realizing their potential. By leveraging microgrid technology, societies can move toward a more sustainable energy future, empowering communities and industries with a resilient and environmentally friendly electricity infrastructure [14].

4.2.4 GSE resiliency

To reduce the effects of climate change, the current fossil-based energy system must transition to GSE system. Resilience is often mentioned as an additional objective or requirement. Despite its frequent use, resilience has different meanings in different contexts. Regarding energy, if the energy system has one clearly defined equilibrium state, then resilience is defined in relation to the response of the energy system to a disturbance and its ability to quickly return to its equilibrium. The second type of resilience allows for different equilibriums, to which a resilient energy system can move after a disruption. Another type of resilience focuses more on the process and the actions of the system in response to disruption [15]. Achieving resilience requires the ability of the system to adapt and change. The growth of IoT and smart grids provides this opportunity.

The 21st century's booming population and escalating energy demands are driving efforts to enhance the Energy Hub (EH). This drive necessarily includes GSE. The goal is to create a more intelligent and responsive system that can effectively meet consumer needs while simultaneously improving the reliability and efficiency of contemporary energy infrastructure.

This embodies resilience. The Internet of Things (IoT) has emerged as a key enabling technology for Smart Energy Hubs (SEH) and Smart grids. While IoT offers a plethora of innovative solutions across various sectors, including critical infrastructure, it also introduces new security challenges. New research emphasizes cutting-edge approaches and tools that can bolster the security and resilience of IoT-enabled SEH/grids against contemporary physical and cyber threats. Implementing secure and advanced data transmission systems based on blockchain technology holds promise in safeguarding the entire EH from cyber-physical attacks in the future [16]. These efforts support the homeostaticity of energy systems: to bring about a rapid, effective, and efficient state of equilibrium between energy supply and expenditure always, whatever the circumstances, to preserve the stability of systems operation [17].

4.3 IMPORTANCE OF METRICS

As was stated in the Prolog to this book, there is a universe of energy available, but the challenge is to draw upon it and use it sustainably, greenly, and resiliently. The main difference between our natural ecosystems and universal energies lies in their forms. The energy-flow accounting in GSE from energetic and green sustainable metrics must use indicators to assess the energy stored in potential sources and the effect on sustainability its use will have [18] (see Figure 4.4).

Green sustainable energy is a component of the sustainability trajectory. Indicators and metrics are necessary to provide at least a semi-quantitative assessment of progress toward or away from sustainability and resiliency. Otherwise, it becomes impossible to objectively assess whether progress

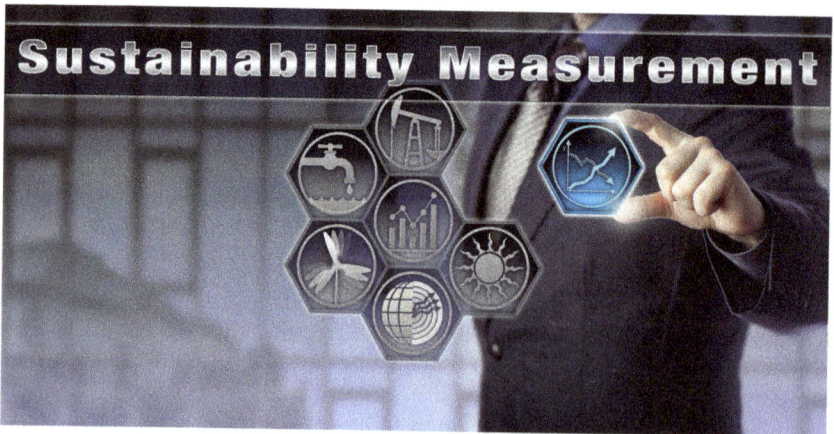

Figure 4.4 Sustainability metrics.

Source: Shutterstock Photo ID: 630330098.

is being made. The subject is by its nature complex and multidisciplinary. Indicators and metrics must, therefore, encompass a wide range of issues relevant to human existence, and they must be useful in "steering" the system toward its goals. They must be well grounded in science and allow for comparisons across different systems. As a minimum, metrics must represent economic strength, human environmental burden, energy use, and system order and stability [19].

To achieve an energy system that is sustainable and resilient, objective metrics to measure GSE are required (Figure 4.3). A set of science-based metrics allows the GSE to increase in use and to make the greatest positive impact [20, 21]. Sustainability metrics (feasibility, viability, desirability) for energy must incorporate resiliency so that sustainable industrialization and economic development will include the economic, environmental, social, cultural, technical, and political dimensions [22].

Carbon and nitrogen footprints are common metrics used to measure the environmental impacts of activities and consumption in society [23]. They can also be used in the transition to GSE. Lowering the energy carbon/nitrogen footprints and enhancing local energy resilience are important. Additional metrics are needed that can assess the impact of energy transition plans and support decision-making to select technologies that create efficient, reliable, and accessible energy systems. These metrics include environmental, technical, social, economic, political, and institutional dimensions. Sustainable energy planning and resiliency are an integrated approach to energy transition planning. Mapping the potential for clean renewable energy at various scales highlights the importance of resilience in energy planning, and addresses challenges associated with energy source selection [24].

4.3.1 Energy metrics

To replace the current energy system, it is imperative to fully describe all its dimensions. Effective Energy Performance Indicators (EEPIs) make it possible to carefully monitor energy performance on a continuous basis, allowing the identification of new ways to optimize and reap the rewards of energy efficiency over time. Effective energy management hinges on the meticulous tracking of specific metrics. These metrics act as a compass, guiding energy managers through the complexities of energy use and its implications. There are seven classes of key metrics for effective energy management (Table 4.1) [25].

Energy management requires tools for real-time tracking and analysis. Advanced metering technologies and Energy Management Software solutions provide a wealth of data, enabling detailed tracking and analysis of energy use. This data is invaluable for making informed decisions that can lead to significant cost and energy savings [8]. It is helpful to establish a framework of energy performance with energy management methods and

Table 4.1 Seven classes of key metrics for effective energy management [24]

Class of metric	Value of metric	Metrics	Description
1. Energy Consumption	• Essential for cost management and ensuring the energy grid's stability	Peak demand	The highest level of energy demand observed in each period
	• Identifying waste and improving overall energy use patterns	Base load energy	The constant minimum level of demand on an electrical system
2. Energy cost	• Shift high-energy processes to times when energy is cheaper	Time-of-use costs	Costs vary by the time when energy is consumed
		Demand charges	Costs incurred based on the highest level of energy demand
3. Energy intensity	• Measure of energy efficiency (energy used/m²) It's a vital indicator.		
4. Carbon footprint metrics	• Understand and manage their environmental impact	Greenhouse gas emissions	Total emissions from a company's operations
5. Renewable energy percentage	• Key indicator of a business's commitment to sustainable energy use		
6. Energy savings metrics	• Energy savings indicate effectiveness of energy management strategies	Efficiency improvements	The quantifiable reductions in energy use due to efficiency measures
7. Return on investment (ROI)	• Justification of investments in energy efficiency and guiding future financial decisions	Payback period	The time it takes for an investment in energy efficiency to pay for itself

energy models in energy monitoring, evaluation, optimization, and benchmarking. As will be seen in Chapter 8, these metrics support the development and use of smart grid technologies.

4.3.2 Exergy

Exergy is an indicator for energy sustainability studies, and it can function as a GSE sustainability indicator [26]. Exergy is a thermodynamic property that links the first and the second thermodynamic principles as well as connects a system under study with the environment where it belongs. Since the first principle of thermodynamics measures the quantity of energy and the second measures irreversibility, exergy combines the two and becomes a great advance in energy sustainability studies. Using exergy for sustainability studies presents some problems, but still offers a worthy contribution to this field, thus complementing traditional economic approaches which are mainly focused on the economic and social poles of sustainability [27].

To evaluate the efficiency of an energy system, both energy and exergy perspectives should be considered, and sustainability is thus measured through exergetic parameters. The energy system's economic feasibility is analyzed based on its fuel costs and operational losses. The results of the study indicate that the energy and exergy efficiencies are low. To measure the sustainability of the system, eight key indicators are used: the depletion number, exergy sustainable index, cumulative exergy loss, relative irreversibility, lack of production, the wasted exergy ratio, environmental effect factor, and improvement potential. The results indicate that the energy system still has opportunities for improvement in terms of sustainability [28].

Given the global population and economic growth, the energy and resource efficiency will become an increasingly competitive factor on the road to sustainability. Among energy efficiency optimization approaches, thermodynamics methodologies contribute toward the improvement of energy efficiency in all energy uses. Besides energy balance, exergy is considered a practical thermodynamics method for the evaluation of any system's energy. From the exergy analysis, merging both exergy efficiency and exergy destruction highlights the energy inefficiencies within a system and provides useful information to the managers and decision-makers for prioritizing the improvement potentials. Exergy analysis is generally an applicable method for the comparison of the alternative processes for a given purpose. There are advantages for the use of exergy analysis methods for current conventional energy efficiency evaluation and for the evaluation of GSE [29].

4.4 METRICS FOR GSE

There is a lack of reliable multidimensional indices to assess GSE adoption across the world. Three metrics were selected: renewable energy output, consumption of renewable energy, and electricity production from renewable sources. These were used to compile the Renewable Energy Index (REI). The REI is a novel measure of adoption of renewable energy sources in countries across the world. This can be applied to GSE (see Figure 4.5). It has the potential to be used in economic analyses and planning for sustainable development. The Index may be useful in comparing countries and regions to encourage investment in green sustainable energy production [6].

It can also be possible to create metrics using multi-criteria decision analysis. This is done by focusing on long-term capacity planning for resource adequacy and sustainability. If stakeholders are concerned about the full

Figure 4.5 GSE metrics.

Source: Shutterstock Photo ID: 2455247321.

range of sustainability metrics – including costs, climate change, pollution, land-use, jobs, and safety alongside water conservation and nuclear concerns – then the various factors can be adjusted to achieve the most sustainable solution [42]. In countries like the United States, which has experienced heightened interest in clean energy sources, new metrics will assist in energy adoption and generation. A challenge with GSE sources (particularly solar and wind) is the inherent variability of energy generation due to climate and geographic influences. Consequently, there is increased interest in addressing energy storage technologies to better capitalize on excess power from high-production periods for use in times of low generation [43]. Energy storage will be discussed in Chapter 8, "GSE harnessing, harvesting, and storage."

Throughout the discussion of new metrics, the importance of computers, the internet, and artificial intelligence (AI) will be clear. Computer-aided methodologies for designing sustainable supply chains will be analyzed using cost, the cost of producing electricity, and two sustainability metrics: the ecological footprint and the energy input. These metrics respectively represent environmental burden in terms of land use and energy resources. New methodologies demonstrate the optimal design of energy systems that provides electric power and heat to global locations. Green sustainable energy resources can yield substantial cost reductions of up to 17%, as well as significant ecological footprint and energy reductions [45].

4.4.1 Lifecycle assessment (LCA)

Life cycle assessment theory, practice, and methodologies are easily applied in GSE design, generation, and use. Providing fundamental knowledge of how to measure sustainability metrics using life cycle assessment in GSE, the inevitable adoption of new energy sources and practices will be facilitated [30]. Life cycle assessment (LCA) and LCA-related approaches represent a powerful method for designing and evaluating the sustainability of green processes. This is mainly due to the intrinsic nature of the LCA method in carrying out the assessment by considering a life cycle perspective, thus including all the involved processes from raw materials extraction to the end of life. The LCA method has been widely adopted for assessing the environmental performance of products or processes in green chemistry and sustainable energy. The LCA method can help to achieve a more comprehensive analysis in a sustainability context when applying it to GSE [31].

An example of the use of LCA as a metric is with hydrogen energy. In the pursuit of sustainable energy solutions, hydrogen is emerging as a promising GSE. The United States has the potential to sell wind energy at a record-low price of 2.5 cents/kWh, making hydrogen production electricity up to four times cheaper than natural gas. With an energy content equivalent to 2.4 kg of methane or 2.8 kg of gasoline per kilogram, hydrogen boasts a superior energy-to-weight ratio compared to fossil fuels. A life cycle assessment (LCA) to evaluate hydrogen's full life cycle, including production, storage, and utilization underscores the importance of measuring hydrogen's environmental sustainability and energy consumption. Key findings reveal diverse hydrogen production pathways, such as blue, green, and purple hydrogen, offering a nuanced understanding of their life cycle inventories. The impact assessment of hydrogen production also reveals environmental implications. Comparative LCA analysis across different pathways provides crucial insights for decision-making, shaping environmental and sustainability considerations. Ultimately, LCA has a pivotal role in guiding the hydrogen economy toward a low-carbon future, positioning hydrogen as a versatile energy carrier with significant potential [32].

Despite the popularity of life cycle assessment (LCA) for understanding the broader implications of technological alternatives, it faces challenges due to difficulties in defining the analysis boundary, uncertain data, and ignoring the role of ecological goods and services. The latter natural capital is essential for sustainability but is only considered partially in conventional LCA. Methods based on mass and energy are possible. Three LCA techniques with different system boundaries: conventional process-based LCA, hybrid economic input-output LCA (EIO-LCA), and hybrid ecologically based LCA (Eco-LCA) can be potential GSE metrics. From LCA analyses, a hierarchical set of metrics can be useful for gaining insight into the role of ecosystem resources in a life cycle. Among aggregation methods, ecological cumulative exergy consumption seems to provide the most meaningful results, but more empirical studies are needed [33].

4.4.2 Computer modeling and AI

Advances in modeling, simulation, optimization, and integration of resource/energy chemical processes in recent years greatly assist both the identification and the measurement of novel GSE. It provides industry and decision-makers with a profound understanding of co-benefits and unintended impacts of the large-scale deployment of various process technologies on the environment and resources in a life cycle perspective. By establishing life cycle models from feedstock, production, market, to recycling, integrated approaches can be explored to evaluate the efficiency and sustainability of alternative processes. Furthermore, these approaches aim to rationalize and optimize flow-sheeting, reduce investment and operating costs, raise efficiency, and minimize environmental impacts. Mass and exergy flow diagram, life cycle inventory, and sustainability indicators can be incorporated into these models. Resource/energy utilization efficiency, environmental impact, and economic benefits can be quantitatively evaluated [34].

GSE will increasingly meet the world's final energy consumption. Hence, seeking ways to measure energy demands and mitigate adverse environmental impacts is necessary. Artificial intelligence (AI) techniques such as machine and deep learning have been increasingly and successfully applied to develop solutions for the energy system. Different AI-based techniques can be employed to resolve interconnected problems related to energy forecasting and management, air quality and occupancy comfort/satisfaction prediction, environmental effects, and effects on sustainability and resilience. AI-based techniques focus on the framework, methodology, and performance. Hundreds of machine learning algorithms can be applied to energy studies. Within the existing and potential programs, there is a wide range of evaluation metrics, adding to the potential applications. While machine and deep learning have been successfully applied in energy efficiency research, most of the studies are still at the experimental or testing stage, and there are limited studies that implemented machine and deep learning strategies in energy applications [35].

4.4.3 Resiliency metrics for GSE

Improving energy performance has been recognized as an effective measure to promote energy saving and emission reduction and to realize sustainable development. When these are coupled with GSE sources, the potential achievement of sustainability is more likely. A contemporary goal for all energy (sources, harvesting, and use) is resilience. Resilience is frequently mentioned as an additional objective or requirement. Here, resilience is defined as the ability of the system to adapt and change. Resilience for the energy system differs depending on which aspect of the energy system is considered, and which elements might be necessary for attaining sustainability [15].

The integration of sustainable energy resources into the current energy system strengthens the resiliency of the entire system. This is the case, especially in case of outages caused by extreme events. GSE provides support to established restoration schemes and increases resilience criteria for a quick recovery of high-priority loads [36]. The bulk electrical system provides a good example of this.

An example of this goal within energy is electrical system resiliency. The bulk electrical system, a critical infrastructure for societal functionality, must meet the electricity demands of end-users sustainably, economically, and in compliance with standards. The concept of electrical system resiliency has gained more importance as natural disasters and cyber threats, and their impacts on power system components could compromise the electrical system. Quantitative metrics from operational and infrastructural perspectives can better measure the progress toward this goal. The integration of smart grid technologies – including demand response, flow of electricity, distributed generation, energy storage systems, and microgrids – are ways to enhance resilience against external shocks. Metrics for resiliency and resiliency strategies are a key way to characterize GSE [37].

In addition to the seven classes of metrics stated in Section 4.3.1 and the concept of resilience, a final metric will be the Internet of Energy (IoE). The quickening growth within the areas of information and communications technology and energy networks has triggered the emergence of a central idea termed IoE. This concept is related to the Internet of Things [38]. The development of an interconnected energy network is being formed through upgrades in the field of intelligent energy systems to control real-time energy optimization and management. This metric will integrate the smart transmission and communication infrastructure, smart metering, pricing, and energy management of energy. The integration, security, and energy management challenges will need to be addressed to ensure higher resiliency, cyber-security, and stability [39].

4.4.4 Metrics for energy efficiency, sustainability, and economic growth

There are two dimensions to "sustainability" as it applies to an energy system: (i) it needs to be designed, operated, and managed such that its environmental impacts and costs are minimal, and (ii) it ought to be designed and configured in such a way that it is robust to extreme disruptions and shocks posed by natural, manmade, or random events, i.e., be resilient under extreme events. These attributes demand a robust set of metrics to ensure their utility. These metrics must include three sustainability indices that monetize the economic, environmental, and resiliency characteristics throughout the lifecycle of the system components. The proposed framework, thus, allows translating sustainability goals into the energy system, and is applicable to:

(i)   the design of new systems,
(ii) assessing performance of an existing system,
(iii) day-to-day scheduling and operation of the system, and
(iv) future growth planning.

A novel presentation method, which can be called "Sustainability Compass," allows decision-makers to visually track the direction and magnitude of changes in the individual sustainability indices of different scenarios; this would also allow easier communication with other stakeholders [40].

4.5 CONCLUSION

There is an increased role of energy in economic development, and this necessitates an improvement of the metrics employed [41]. Previous discussions on Sustainable Development Goal-7 (Affordable and Clean Energy) and Sustainable Development Goal-13 (Climate Action) emphasize the value of metrics to evaluate GSE [42]. Through the use of exergy as a measure for identifying and explaining the benefits of sustainable energy and technologies, benefits can be clearly understood, and the utilization of green sustainable energy and technologies can be increased. Exergy and resiliency can be used to assess and improve energy systems and can help better understand the benefits of utilizing green sustainable energy by providing more useful and meaningful information. These metrics clearly identify efficiency improvements and reductions in thermodynamic losses attributable to more sustainable technologies. As a new sustainability index, exergy can also identify better than prior energy measurements the environmental benefits and economics of energy technologies. GSE metrics should be utilized by engineers and scientists, as well as decision- and policymakers, involved in green energy and technologies in tandem with other objectives and constraints [43].

REFERENCES

1. Hare, T., Measuring up to nature. *Ecos*, 2007. **28**(2): p. 20–24.
2. Ceratti, D.R., *et al.*, Response to comment on "Eppur si Muove: Proton diffusion in halide perovskite single crystals": Measure what is measurable, and make measurable what is not so: Discrepancies between proton diffusion in halide perovskite single crystals and thin films. *Advanced Materials*, 2021. **33**(35).
3. Andraos, J., Aiming for a standardized protocol for preparing a process green synthesis report and for ranking multiple synthesis plans to a common target product. *Green Processing and Synthesis,* 2019. **8**(1): p. 787–801.
4. Redick, T.P., Sustainability standards and metrics: Reach for ANSI-ISO? *SESHA Journal: Semiconductor Environmental Safety and Health Association*, 2010. **210**.

5. Danish, M.S.S., *et al.*, Energy and environment efficiencies towards contributing to global sustainability, in *Sustainability Outreach in Developing Countries*. 2020, Springer Singapore. p. 1–13.

6. Outili, N. and A.H. Meniai, Green chemistry metrics for environmental friendly processes: Application to biodiesel production using cooking oil, in *Nanotechnology in the Life Sciences*. 2020, Springer Science and Business Media B.V. p. 63–95.

7. Hemmati, M., N. Bayati, and T. Ebel, Integrated life cycle sustainability assessment with future energy mix: A review of methodologies for evaluating the sustainability of multiple power generation technologies development. *Renewable Energy Focus*, 2024. **49**.

8. Hoang, P., *et al.*, What metrics to evaluate sustainability of photovoltaic systems? *Metallurgical Research and Technology*, 2014. **111**(3): p. 201–210.

9. Muniz, R.N., *et al.*, The sustainability concept: A review focusing on energy. *Sustainability (Switzerland)*, 2023. **15**(19).

10. Gan, V.J.L., *et al.*, Simulation optimisation towards energy efficient green buildings: Current status and future trends. *Journal of Cleaner Production*, 2020. **254**.

11. Vogt, K.A., *et al.*, Bioresource-based energy for sustainable societies, in *Handbook of Sustainable Energy*. 2011, Nova Science Publishers, Inc. p. 89–137.

12. Younesi, A., *et al.*, Trends in modern power systems resilience: State-of-the-art review. *Renewable and Sustainable Energy Reviews*, 2022. **162**.

13. Ren, H., *et al.*, Optimal planning of an economic and resilient district integrated energy system considering renewable energy uncertainty and demand response under natural disasters. *Energy*, 2023. **277**.

14. Us Salam, I., *et al.*, Addressing the challenge of climate change: The role of microgrids in fostering a sustainable future: A comprehensive review. *Renewable Energy Focus*, 2024. **48**.

15. Jesse, B.J., G.J. Kramer, and V. Koning, Characterization of necessary elements for a definition of resilience for the energy system. *Energy, Sustainability and Society*, 2024. **14**(1).

16. El-Afifi, M.I., *et al.*, A review of IoT-enabled smart energy hub systems: Rising, applications, challenges, and future prospects. *Renewable Energy Focus*, 2024. **51**.

17. Yanine, F., *et al.*, Integrating green energy into the grid: How to engineer energy homeostaticity, flexibility and resiliency in electric power distribution systems and why should electric utilities care, in *Low Carbon Energy Technologies in Sustainable Energy Systems*. 2021, Elsevier. p. 253–266.

18. Marull, J., *et al.*, Using thermodynamics to understand the links between energy, information, structure and biodiversity in a human-transformed landscape. *Ecological Modelling*, 2023. **476**.

19. Cabezas, H., Sustainability indicators and metrics, in *Sustainability: Multi-Disciplinary Perspectives*. 2012, Bentham Science Publishers Ltd. p. 197–221.

20. Buonocore, J.J., *et al.*, Correction: Metrics for the sustainable development goals: Renewable energy and transportation. *Palgrave Communications*, 2019. **5**(1): p. 136.

21. Buonocore, J.J., *et al.*, Metrics for the sustainable development goals: Renewable energy and transportation. *Palgrave Communications*, 2019. **5**(1).

22. Okorie, M.E., *et al.*, Renewable energy and African industrialization: A participatory integrated approach in assessing concentrated solar power potential in Namibia. *International Journal of Mechanical Engineering and Technology*, 2018. 9: p. 509–524.

23. Natyzak, J.L., *et al.*, Virtual water as a metric for institutional sustainability. *Sustainability (United States)*, 2017. 10(4): p. 237–245.

24. Wehbi, H., Powering the Future: An Integrated Framework for Clean Renewable Energy Transition. *Sustainability (Switzerland)*, 2024. 16(13).

25. EnergyAction. 7 key metrics for effective energy management. *Energy Insights*, 2024; Available from: https://energyaction.com.au/metrics-for-effective-energy-management-australia/.

26. Hai, T., *et al.*, Applying energy-exergy, environmental, sustainability, and exergoeconomic metrics and bi-objective optimization for assessment of an innovative tri-generation system. *International Journal of Hydrogen Energy*, 2024. 52: p. 315–333.

27. Romero, J.C. and P. Linares, Exergy as a global energy sustainability indicator: A review of the state of the art. *Renewable and Sustainable Energy Reviews*, 2014. 33: p. 427–442.

28. Maruf, M.H., *et al.*, Energy and exergy-based efficiency, sustainability and economic assessment towards improved energy management of a thermal power plant: A case study. *Sustainability (Switzerland)*, 2023. 15(6).

29. Taheri, K., R. Gadow, and A. Killinger. Exergy analysis as a developed concept of energy efficiency optimized processes: The case of thermal spray processes. in *Procedia CIRP*. 2014. Elsevier B.V.

30. Mahmud, M.A.P., *et al.*, *Green Energy: A Sustainable Future. Green Energy: A Sustainable Future*. 2023, Elsevier. p. 1–239.

31. Mondello, G. and R. Salomone, Assessing green processes through life cycle assessment and other LCA-related methods, in *Studies in Surface Science and Catalysis*. 2019, Elsevier Inc. p. 159–185.

32. Osman, A.I., *et al.*, Life cycle assessment of hydrogen production, storage, and utilization toward sustainability. *Wiley Interdisciplinary Reviews: Energy and Environment*, 2024. 13(3).

33. Urban, R.A. and B.R. Bakshi, 1,3-Propanediol from fossils versus biomass: A life cycle evaluation of emissions and ecological resources. *Industrial and Engineering Chemistry Research*, 2009. 48(17): p. 8068–8082.

34. Qian, Y., *et al.*, Life cycle assessment and sustainability of energy and chemical processes. *Huagong Xuebao/CIESC Journal*, 2013. 64(1): p. 133–147.

35. Tien, P.W., *et al.*, Machine learning and deep learning methods for enhancing building energy efficiency and indoor environmental quality: A review. *Energy and AI*, 2022. 10.

36. Qin, C., *et al.*, An integrated situational awareness tool for resilience-driven restoration with sustainable energy resources. *IEEE Transactions on Sustainable Energy*, 2023. 14(2): p. 1099–1111.

37. Erenoğlu, A.K., I. Sengor, and O. Erdinç, Power system resiliency: A comprehensive overview from implementation aspects and innovative concepts. *Energy Nexus*, 2024. 15.

38. Nelson, W.M., Sustainable agricultural chemistry in the 21st century: Green chemistry nexus, in *Sustainable Agricultural Chemistry in the 21st Century: Green Chemistry Nexus*. 2023, CRC Press. p. 1–294.

39. Farhan, M., *et al.*, Towards next generation internet of energy system: Framework and trends. *Energy and AI*, 2023. **14**.

40. Moslehi, S. and T.A. Reddy, A new quantitative life cycle sustainability assessment framework: Application to integrated energy systems. *Applied Energy*, 2019. **239**: p. 482–493.

41. Ayres, R.U., H. Turton, and T. Casten, Energy efficiency, sustainability and economic growth. *Energy*, 2007. **32**(5): p. 634–648.

42. LaBelle, M.C. and T. Szép, Green economy: Energy, environment, and sustainability, in *Contributions to Economics*. 2022, Springer Science and Business Media Deutschland GmbH. p. 325–364.

43. Rosen, M.A., I. Dincer, and M. Kanoglu, Role of exergy in increasing efficiency and sustainability and reducing environmental impact. *Energy Policy*, 2008. **36**(1): p. 128–137.

Chapter 5

Politics and economics of green sustainable energy

Reformers have the idea that change can be achieved by brute sanity.

– George Bernard Shaw [1]

5.1 INTRODUCTION

We are experiencing a global energy system transformation (GEST) which is altering the world's economy, opening new axes of political unrest, and revolutionizing the energetic basis of human civilization (Figure 5.1). Energy geopolitics and economics have not fully addressed this new reality. How will geopolitics and economics direct the adoption of GSE? There are three vital areas of geopolitical economies that will influence the process: the wide-ranging material dimensions of the transformation, its geographical extension, and its conflict-ridden political economy. The adoption of new energies is inevitable, so it becomes necessary to examine the roles geopolitics and economics play in its development [2].

Figure 5.1 The geopolitics and worldwide economics of GSE.

Source: Shutterstock Illustration ID: 2297429939.

DOI: 10.1201/9781003407447-7

The transition to green sustainable energy is a gradual, complicated, and challenging process that involves multiple stakeholders and dimensions. Therefore, it is necessary to identify the key aspects and features of this complicated process. This analysis will be of great interest as it has important implications for GSE acceptance. It is important to realize that while technical factors are prominent, the economic, institutional, social, and even psychological aspects of the transition to renewable energies should not be neglected [3].

The PESTEL analysis on dimensions that affect the environment (see Chapter 2, "PESTEL analyses of GSE") [4] is essential in understanding the success or failure of GSE. This chapter will highlight the dynamic effects that geopolitics and green finance have in the development and acceptance of green sustainable energy (GSE). Causal relations between geopolitical risk and green finance are strong and critical. Importantly, geopolitical risk has a prolonged impact on the volatility of green financing and sustainability. Yet, geopolitical risk also tends to influence the return to the use of clean energy more persistently. Global geopolitical risk underscores the necessity of promoting the development of GSE to reduce the dependence on fossil fuels and enhance energy independence [5].

5.2 IMPORTANCE OF GEOPOLITICS AND ECONOMICS

In an era defined by unprecedented industrialization and technological advancement, the world faces an increasingly intricate web of energy challenges that affect economies, societies, and international relations. As governments, industries, and societies grapple with the complexities of energy demand, security, economics, and sustainability, it is clear that geopolitics and economics are vital to GSE transition (Figure 5.2). Energy crises have both supply-side vulnerabilities and demand-side pressures, and these are important. Major energy-producing nations and organizations shaping global energy markets play a pivotal role in deciding between power and resources. Under these conditions, it is critical to comprehend the challenges at hand and to champion transformative solutions that are necessary to position our world for sustainability and resiliency in terms of energy [6].

It is helpful to look at how natural resource dependence (NRD) and natural resource abundance (NRA) play a role in energy and environmental concerns (Figure 5.1). This also influences geopolitics and economics. Fiscal decentralization is beneficial for environmental sustainability, especially across countries with a higher level of energy and carbon intensity. However, enhanced financial inclusivity is detrimental to environmental quality, as occurs in more energy-efficient economies. As per the direct effects, NRD and NRA have effects in all societies. Concerning the indirect effects, NRD and NRA have more substantial effects in more energy-efficient economies.

Figure 5.2 Balancing investment decisions in uncertain world weighing opportunities.

Source: Shutterstock AI-generated image ID: 2452156519.

Environmental innovation, renewable electricity, employment-to-population ratio, and economic progress enhance environmental sustainability. Governmental oversight further enhances environmental protection. Moreover, finances can enhance the access to and affordability of financial services to economic agents for green consumption and investment ventures to achieve environmental sustainability, among other Sustainable Development Goals [7].

5.2.1 Global policy and politics

The world's largest developing countries, the world's largest manufacturers, and countries with highly developed industrial economies must grapple with multiple challenges such as climate change, economic transition, and environmental protection. These challenges will require systematic support from policy, economy, technology, and society. How these challenges are met will be the determining factor in sustainable development. In the area of energy transformation, which includes industrial structure upgrading, transportation carbon reduction, urban planning and construction, green finance, and supporting policies, politics, and finances are key. Policy planning and fiscal commitment are also vital [8].

Globalization and Politics bring insight and are necessary to address issues. The globalization of energy, economy, environment, and politics shows how they interact and work harmoniously. Providing opportunities for interdependency and governance, globalization offers both dangers and promises when moving toward sustainable development and resilience. Globalization offers an economic opportunity with wide-ranging effects and also has a political bent with cultural implications. These are some of the key challenges for globalization in the 21st century as it is applied to the necessity of GSE [9].

To combat climate change, a country must invest in the development of GSE sources and the upgrading of its energy infrastructure. Political support with capital is necessary to adequately fund the energy revolution because green energy development is often expensive and risky, especially in its infancy. It is imperative to understand how geopolitics and financial development affect GSE consumption as well as which aspects of financial development are most essential to its development. All three facets of financial development (financial development overall, financial market-related development, and financial institution-related development) positively impact green sustainable energy consumption. While all three aspects of financial development increase the adoption of GSE, the effect is stronger in the development of financial markets. This, in turn, will have policy implications for the achievement of Sustainable Development Goals [10].

5.2.1.1 Green growth and degrowth

Green growth is the growth of economies using modern, environmentally sound, energy-efficient technologies, including the use of alternative energy sources. Degrowth, on the other hand, critiques the global capitalist system which pursues growth at all costs and prioritizes social and ecological well-being instead of corporate profits, over-production, and excess consumption. Green growth and degrowth are valuable topics on socio-ecological transformation. Green growth is highly policy oriented, focusing on practical implementation, and builds on the actual use of green technologies. Degrowth focuses on human-nature interrelationships and human responses to overgrowth [11]. Social change and transition utilize these concepts, and they contribute to the future adoption of GSE.

"Green growth" has received international attention, and it means this global endeavor can continue without surrendering continued economic growth by a redirection of human effort to invent green energy technology, green energy markets, and green energy choices. This green energy-based economic growth represents a paradigm shift bringing sustainable and equitable relations between environment, economy, and society [12].

Sustainable consumption and production (SCP) underlies the development, harnessing, and use of GSE. The present status and existing trends of SCP have behaviors that align with green growth goals and necessitate the application of sustainable growth or degrowth. The future development of SCP will also favor degrowth and it will require reduced personal and collective consumption [13]. Sustainable degrowth offers effective alternative strategies for tackling social and environmental problems such as climate crisis, resource depletion, and transitioning to GSE. However, it plays a marginal role in policy. This is because the law can be understood as the universal basis of public policies and fundamental language to express these policies [14]. As understood here, degrowth is an action in response to overgrowth.

Sustainable development in the UN Sustainable Development Goals (SDGs) [15] must find a balance among political, ecological, and economic systems when it comes to development. "Circular Economy," "Degrowth," and "Green Growth" have made important contributions to research related to the 17 UN SDGs [16]. For example, degrowth is a radical political and economic reorganization leading to reduced resource and energy use. It lowers the cost of growth-based development. There are limits to growth due to historical, cultural, social, and political forces that have gone beyond economic growth [17]. This supports the spirit of the SDG, especially SDG 7 [18].

5.2.1.2 Circular resource policy

The green transition requires GSE resources, as it promotes resource circularity if sustainable substrates are used. While subsidies have played a key role in pursuing economic sustainability, their use should be reduced over time and measured to the actual contribution related to environmental and social improvement. A development plan based on the circularity of resources includes subsidies for small-scale plants, substrates from neighboring territories, citizen involvement in decision-making processes, and stability of political choices [19].

There are challenges to building a GSE economy. The development of a Circular Economy (CE), where resources are kept in circulation for the extraction of maximum value, needs both policy protection and economic support. Circular economy-relevant policies involving energy remove inadvertent barriers to resource recovery. Implementing reforms to earlier regulations necessitates changes to previous institutional practices. Circular economy theory and policy need to be aware of policy legacy. Policy needs to go beyond the short-term economic concerns and to prioritize reuse, refurbishment, or recovery of value (via extended producer responsibility). Governments must establish agencies charged with resource management, stewardship, and productivity if the purported benefits of CE are to be realized [20, 21]. How this can be applied to GSE will become clear over time.

5.2.1.3 International commitment to energy issues

Creating an effective energy policy is hard, in part because it often requires effective international coordination. Solutions often lie in making problems more manageable by working in small groups of relevant countries; successful cooperation also hinges on finding incentive-compatible commitments that align, to the extent feasible, with national interests and are focused on areas where cooperation will yield tangible joint gains [22].

It is possible to identify potential energy security risks and opportunities of new renewable energy (RE) export projects that will garner transnational interest and will promote GSE. These opportunities will necessarily include discussions of human security, geopolitical/foreign policy, and material risks. Inevitably, questions arise regarding price stability, affordability, access and equity; reliability and resilience risks; and environmental impacts. Ultimately, all of these considerations can inform the development of new transnational GSE projects, and improve forward-thinking on risk management [23].

Increasing international commitment is required to harness research to contribute to solving grand societal challenges related to environmental change. Examples include global research programs like the United Nations Sustainable Development Goals [24]. On a science–policy level, funders, policymakers, and scientists must come together to discuss how to overcome the key obstacles in the path of such change [25]. Globalization and international commitment will involve the cooperation of societies and economies (and all their institutions including educational and scientific). This will encourage all countries to form a single socioeconomic system with the prospect of more effective political coordination among them. This will necessitate a new type of social contract between society and citizens, which must necessarily be included in modern institutional systems. This contract will be formed from different national cultures, so the socioeconomic development should be based on intercultural dialogue between different countries, social systems, and civic institutions [26].

As has been presented in this book, it has been more than five years since the adoption of the 2030 Agenda for Sustainable Development with its Sustainable Development Goals (SDGs) and the Paris Agreement, which seek to improve the well-being of people and the planet and strengthen the global response to the threat of climate change. While these major international commitments have spurred a lot of policy debates and academic research, a synthesis of how their adoption has shaped the academic discussions in pursuit of these goals in specific sectors such as energy is limited. The three most common policy-related issues identified are a lack of integrated/cross-sectoral planning, a narrow emphasis on energy justice in policies, and the need for more cost-effective strategies in pursuit of the Paris Agreement. There will always be a need for research on the progress of implementation, impacts, and critical lessons from current policy efforts to achieve these global agendas [27].

5.2.1.4 Role of geopolitics and economics in energy transition

There is a dynamic relationship between geopolitics and global economics, affecting the growth of and transition to GSE in modern global international society. The international political system influences both logics and dynamics, separate from those of modern capitalism. Geopolitics and global economics have changed in fundamental ways and it is more than a competition between Western dominance or the rise of China. The identification of some of the most important institutional pinch-points where geopolitics and global economics intersect include sanctions, security-related trade measures, and industrial strategy [28].

Countries with high vulnerability to climate change are more likely to experience geopolitical conflicts. Country-level overall economic, social, and governance readiness significantly mitigates this detrimental effect [29]. Addressing and adopting GSE is crucial to promoting global peace and geopolitical stability. Geopolitical risk potentially impacts economic, political, social, and environmental aspects on the stability and security of a country or region. This risk can range from minor disruptions to major conflicts that can significantly affect the global economy and may affect countries in various ways, including disruptions to supply chains, demographic shifts, cultural differences, social and climate change, natural disasters, and resource depletion [30]. As energy availability becomes more important, geopolitics and economics will play greater roles.

Energy colonialism is essential for understanding how past, present, and future energy systems are shaped by colonial or neocolonial power dynamics, imaginaries, discourses, and practices. This can shed light on current energy transition processes, namely with regard to green finance flows, new green geopolitics, and energy governance. Energy colonialism is shown when power/influence is exerted over energy transition processes. It is felt in scientific development and knowledge transfer, but also as an intervention on an individual scale, affecting daily life and human-nature relations. Colonial continuities are pervading contemporary energy debate:

- green hydrogen production in the Global South to sustain economic growth in the Global North;
- in energy partnerships; and
- in financial dependencies that stabilize the political economy of clean energy.

A more nuanced understanding of energy colonialism concept may serve as a multidimensional research strategy for critical social science research on energy transitions and provide guides for GSE development and infrastructures, and further broad acceptance and adoption [31].

5.2.1.5 Key to sustainable development

Economic development has brought some phenomenal benefits, such as rising life expectancy and improved overall public health. It has also had some planet-threatening adverse effects, such as massive tropical deforestation, ocean fisheries depletion, man-made climate change, and violent competition over limited hydrocarbon resources. Hydrocarbon scarcities can easily lead to war unless some alternatives are developed, including the development of GSE. Global market forces can be "re-engineered" to channel economic activity in a sustainable manner and new approaches to global politics and governance itself, based firmly on the building science of sustainability, can provide a vital bridge to future prosperity and peace [32].

The sustainability of an institution, organization, or a program like GSE requires a constant review of strategic positioning and the execution of pertinent plans in response to evolving external demands. Resilient organizations continue to revive themselves through effective R&D and the renewal of their range of products and services. Financial and technological innovation for sustainability must include environmental, social, and governance dimensions. GSE must examine approaches to sustainability under the ongoing development of energy sustainability and green finance initiatives. The intertwined public-private partnership and implications of geopolitics under an evolving global financial system for sustainability transformation are highly sought [33].

The topic of green growth is relevant here. In a modern, dynamically changing society, GSE sources are of interest to an increasing number of people and to growth of economies. When discussing emerging problems, geopolitical and economic organizations conclude that the key to the solution of the issues lies in the use of environmentally friendly renewable energy sources. The development of GSE solidly addresses concerns in both the fuel and energy sectors, which supports green growth [34]. The transition to sustainable development and resiliency is impossible without reducing the impact of energy on the environment, while simultaneously meeting global energy demands.

5.3 GEOPOLITICS IN THE TRANSITION TO GSE

Geopolitical risks and uncertainty negatively affect the economy and environmental quality. High geopolitical risks (risks include energy, climate change, nationalism, protectionism, and cyberattacks) and high uncertainty increase carbon emission intensity by increasing dependence on fossil fuels (see Figure 5.3). This, in turn, diminishes seeking GSE. Environmental and energy policymakers should mitigate carbon intensity through long-term rather than short-term mechanisms to achieve SDG12 and SDG13 [35]. In order for energy transition to lead to sustainable development, geopolitical drivers

Figure 5.3 New energy geopolitics: global transition and trade.

Source: Shutterstock AI-generated image ID: 2484703085.

must be acknowledged and addressed. Policymakers can use political tools and mechanisms to improve the energy transition and implement a plan to expand the use of sustainable green energy [36].

Geopolitics plays an important role in the development, harvesting, and use of GSE. Green sustainable energy has many advantages over fossil fuels for international conduct, security, and peace. This new energy is thought to exacerbate security risks and geopolitical tensions related to critical materials and cybersecurity; former hydrocarbon exporters will likely be the greatest losers from the energy transition. This misunderstanding may result from a failure to completely define "geopolitics" which leads to an underestimation of the breadth of its influence. There are many types of GSE, and there will be different geopolitical risks associated with each. Consideration must be made on availability of critical materials and cybersecurity, while also raising concerns about the replacement of the existing supply chains and markets. Geopolitical stability will require weighing the risks of transition to new energies, while steadily meeting the existing demands for energy [37]. While there are many geopolitical concerns, energy is one of the most critical [38].

5.3.1 Encouraging growth of GSE

Green sustainable energy resource development, natural resource extraction, and the possibility of green economic growth are all signs of successful acceptance of GSE. New energy resources can stimulate the economy while also promoting environmental sustainability. To work toward a more sustainable world, regulations on GSE need to consider environmental impact, minimize inconsistencies, and encourage environmental education [39]. There is a direct relationship between green sustainable energy and the peaceful execution of geopolitics. Energy projects are strategically sited as part of, near, across, and even distant from political borders; this will mean that these projects will involve the countries' economic and labor resources. Competing relationships will exist between the physical, political, and regulatory conditions prevailing in neighboring countries and the development of GSE [40].

Conflicts arise and can increase the risk associated with new energy projects due to disruptions in the energy supply chain. Energy uncertainty, geopolitical conflict, and military activity involving use of GSE and nonrenewable energy reveal that energy uncertainty increases GSE consumption and lowers nonrenewable energy consumption. The current geopolitical landscape poses unfavorable conditions for both the expansion of renewable energy usage and the utilization of non-renewable energy sources. While international trade positively influences the growth of GSE, transparent economic strategies are essential for governments to reduce uncertainty and facilitate a transition to GSE [41].

5.3.2 Negative influence on GSE

Geopolitical threats (war threats, peace threats, military buildups, nuclear threats, and terror threats), geopolitical acts (beginning of war, escalation of war and terror acts), and geopolitical risks all impact GSE. Green technologies, natural resource rents, and trade openness also impact energy transition within geopolitical domains. Geopolitical threats, geopolitical acts, and geopolitical risks adversely affect the transition to GSE. Geopolitical threats have a greater hindrance to the energy transition than geopolitical acts. Rising geopolitical risk discourages green sustainable energy transition, although the inhibitory effect diminishes over time [42].

5.3.3 Energy geopolitics

Dramatic technological advances in GSE sources and environmental concerns have set in motion a global energy transformation that is expected to have profound geopolitical consequences. While the current energy prices crisis and the global conflicts threaten to reverse or slow this trend, acceptance of climate change is challenging many countries that already face the reverberations of changes in weather patterns. According to this new

Figure 5.4 Cost of energy.

Source: Shutterstock Photo ID: 2154222979.

approach, geopolitics, geoeconomics, and regional and national politics need to be strengthened by an understanding of the fundamental ecological independences that rule the Earth's systems. These dependencies are vital to understanding today's natural disruptions and the critical importance of ecological integrity for global security [43].

5.4 EFFECT OF ECONOMICS ON GSE

There is a close but complex relationship between financial development (FD), financial risk, green finance, and innovation related to GSE transition (see Figure 5.4). Increasing economic development increases emissions and negatively impacts the environment. Efficient resource allocation, improved financial systems, and green innovation are likely to contribute to emission mitigation and the overall development of a sustainable viable economy. Risk management tempers financial systems when funding is considered for future emissions prevention. Geopolitical effects from implementing green energy policy may be overshadowed by the cost of resource allocation efficiency and technological innovation. GSE policies must spur increased total productivity of small businesses, large-scale firms, and firms with high equity concentration [44]. When this occurs, the likelihood of GSE adoption increases.

5.4.1 Positive effects of economics on GSE

Green sustainable energy and green patents/grants positively impact environmental sustainability and they seem to be strengthened by the circular economy. This may help policymakers to promote ecological sustainability and push them toward the attainment of sustainable development goals [45].

The effects of economics (public funding, grants, and awards) drive research and development (R&D) networks. Core energy technology fields, bolstered by public funding, are critical for enhancing competitiveness and sustainable growth at the nationally strategic technology level. Thus, the relationship between R&D and the level of government funding for these fields is generally perceived as both desirable and necessary. This fact indicates that government-funded R&D should be designed and managed not only to curb the inefficiencies existing in the current funding programs but also to achieve support for further technological development.

5.4.2 Negative effects of economics on GSE

The anthropogenic impact of conventional energy sources encourages the utilization of GSE. It has benefits for economic growth. On the other hand, institutional stability is the prerequisite for environmental quality and a healthy functioning economy. Consumption of both nonrenewable and green sustainable energy maintains economic growth but also causes environmental degradation. GSE, at least in the short term, will affect institutional stability. It has been recommended that it is preferable to have a blend of both types of energy and a gradual transition toward GSE sources, with better implementation of policies and technological advances, to produce, preserve, and transmit energy production [46].

Using fossil fuels alone has a detrimental effect on a nation's ability to grow sustainably. Consequently, the focus of policymakers and environmentalists worldwide has shifted toward green sustainable energy and environmental technologies, which demand more green investments in society. The practice of resource renting hurts green investment. (The resource rent of a natural resource is the total revenue that can be generated from the extraction of the natural resource, less the cost of extracting the resource (including a normal return on investment to the extractive enterprise).) Conversely, political stability, government effectiveness, control of corruption, the rule of law, accountability, and regulatory quality help green investment to increase. Resource rent and different measures of institutional quality hurt green investment. GDP, carbon emissions, and trade openness boost green investment, while the foreign direct investment and consumer price index hurt green investment. Therefore, to promote green investment, policymakers should focus on strengthening domestic institutional quality and reducing reliance on resource rents earned from fossil fuels [47].

5.4.3 Cost of transitioning to GSE

Transitioning the global energy system to GSE will require large-scale investment flows, with a central role needed for transnational climate finance to include private funds. There must be a willingness to provide international financing in accordance with common but differentiated responsibilities (acknowledged by the broad endorsement of the Paris Agreement), and the Green Climate Funds in particular that will facilitate the growth of GSE. It is estimated that the international community must mobilize at least USD 100 billion per year for mitigation and adaptation in developing countries. The support for international climate cooperation is improved when efforts of successively rising domestic energy cost levels are compensated. The global goal of cost-efficient mitigation of environmental areas with national policy priorities (climate finance) could become a central pillar of sustainable development and promote international cooperation to achieve the climate targets laid down in the Paris Agreement [48].

5.4.4 Green and sustainable finance

"Green finance" has evolved due to a growing global concern toward environment protection and climate change mitigation. Green finance and sustainable finance are types of financial activities that support the transition to a low-carbon, sustainable economy while addressing global challenges, such as climate change and emerging environmental and sustainability risks. Green finance involves financing projects and initiatives that have positive environmental impacts such as reducing greenhouse gas emissions and promoting green sustainable energy. In tandem, sustainable finance integrates environmental, social, and governance (ESG) factors into investment decisions to promote long-term economic growth, social outcomes, and environmental sustainability. Both green finance and sustainable finance aim to drive positive change by mobilizing capital toward activities that promote sustainability and reduce negative environmental impacts.

Green finance and sustainable finance are important tools for achieving the transition to GSE by redirecting investments toward environmentally friendly projects and integrating ESG factors into investment decision-making. By incentivizing investments in green sustainable energy, energy efficiency, and other sustainable initiatives, green finance and sustainable finance can help reduce greenhouse gas emissions, mitigate the negative impacts of climate change, and help achieve a sustainable and resilient global economy that promotes long-term social and environmental well-being. Green and sustainable finance is crucial for advancing a GSE system and economic growth. Funding in this area will include areas like urban total factor energy efficiency, reducing carbon emissions, regional economic development, climate adaptation, technological progress, green innovation, and transitioning from linear to circular economy models. Future work will focus on refining industrial and financial structures that align sustainable, green development with superior ecological environment protection, highlighting the evolving nature of green finance research and its increasing impact on promoting sustainable economic practices [49].

The relationship among financial practices, ecological well-being, and economic resilience has begun to receive more attention when environmental sustainability and economic stability are significant issues. The coupling and coordination between green finance and economic resilience are the foundation of sustainable economic development [50]. This further highlights the pivotal role of green finance in fostering economic resilience, particularly in instances of resource scarcity or abundance. Moreover, it is important to maximize the positive impact of green finance on economic systems, especially in the context of natural resource constraints [51].

5.4.5 Green investment

The collateral effects of climate change require the exploration and implementation of appropriate ways to reduce ecological issues while simultaneously maintaining economic and social well-being. Green investing seeks to support business practices that have a favorable impact on the natural environment. Often grouped with socially responsible investing (SRI) or environmental, social, and governance criteria, green investments focus on companies or projects committed to the conservation of natural resources, pollution reduction, or other environmentally conscious business practices. The expansion of GSE allows for a reduction in the negative anthropogenic impact on the environment without restricting economic growth or social welfare. However, the expansion of GSE necessitates additional green investment [52].

In 2022 Oxford University predicted that a rapid switch to green energy sources like hydrogen could save the world an estimated $12 trillion. If that's right, it could significantly accelerate business and government investment in this area [53]. Green finance (GF) and GSE technology innovation will play a pivotal role in paving the way toward the adoption of GSE. They hold immense potential to drive meaningful change and ensure a greener future for generations to come [54]. Countries must aim to increase the share of green sustainable energy sources in energy consumption, as significant measures toward sustainable development. However, several factors affect the successful green energy adoption, where financial considerations remain key determinants in decision-making by companies, families, and institutions. Investing in GSE can be profitable in both the near and long term [55].

5.4.6 Cost of sustaining the earth

In a world of finite resources and limited ecosystem capacity, the prevailing model of economic growth, founded on ever-increasing consumption of resources and emission pollutants, cannot be sustained. In this context, the "green economy" offers the opportunity to change the way that society manages the dynamics of the environmental and economic domains.

To enable society to build and sustain a green economy, innovations (e.g., "green nanotechnology") must exploit discoveries in materials science and engineering to generate products and processes that are energy efficient as well as economically and environmentally sustainable. These applications are expected to impact a large range of economic sectors, such as energy production and storage, as well as construction and related infrastructure industries. These solutions may offer the opportunities to reduce pressure on raw materials trading in GSE, to improve power delivery systems to be more reliable, efficient, and safe as well as to use unconventional water sources or nano-enabled construction products therefore providing better ecosystem and livelihood conditions [49].

5.5 ROLE OF PUBLIC SECTOR

The participation of citizens and communities in the energy transition, which encourages a bottom-up approach to the implementation of sustainable energy initiatives, is essential (Figure 5.5). This is in tune with the United Nations' Sustainable Development Goals, which attempt to involve all members of society in the sustainability path. The reality in many countries, however, is that community GSE still lacks the necessary regulatory framework to compete with large utility companies. This may indicate that the governance framework needs more momentum since it is still not ready to include communities (collective citizens) as full participants in the energy transition [42].

Since energy is an indispensable part of modern society and can serve as one of the most important indicators of socioeconomic development, the active involvement of humans (developers, harvesters, and users) is vital. Whether energy needs are met through traditional means by burning biomass resources (i.e., firewood, crop residues, and animal dung), in crude traditional stoves, or through more modern methods, we must strive to advance to GSE. To achieve sustainable global development it is imperative that GSE is developed, harnessed, and used. Upgrading existing energy resources to cleaner and more efficient energy has unique potential to provide clean and reliable energy, while simultaneously preserving the local

Figure 5.5 Human role in sustainability.

Source: Shutterstock Vector ID: 2467829013.

and global environment. In spite of its significant potential to serve all nations, however, the high costs of installation and maintenance of these systems preclude widespread global adoption. Concerted efforts from both geopolitical and economic sectors are absolutely essential in facilitating modernization and dissemination of GSE technology to harness the inherent potential that is currently underutilized and unexploited [41].

REFERENCES

1. Shaw, G.B. 2024. Available from: https://www.azquotes.com/quote/604131.
2. Kuzemko, C., *et al.*, *Rethinking Energy Geopolitics: Towards a Geopolitical Economy of Global Energy Transformation.* 2024: Geopolitics.
3. Omri, E., N. Chtourou, and D. Bazin, Technological, economic, institutional, and psychosocial aspects of the transition to renewable energies: A critical literature review of a multidimensional process. *Renewable Energy Focus,* 2022. **43**: p. 37–49.
4. Nelson, W.M., Sustainable agricultural chemistry in the 21st century: Green chemistry nexus. *Sustainable Agricultural Chemistry in the 21st Century: Green Chemistry Nexus.* 2023: CRC Press. 1–294.
5. Zhang, D., *et al.*, The causal relationship between green finance and geopolitical risk: Implications for environmental management. *Journal of Environmental Management,* 2023. **327**.
6. Ozcan, M.S.O., Analyzing energy crises and the impact of country policies on the world, in *Analyzing Energy Crises and the Impact of Country Policies on the World*, Merve Suna Özel Özcan, editor. 2023: IGI Global. 1–283.
7. Ahmad, M. and E. Satrovic, Relating fiscal decentralization and financial inclusion to environmental sustainability: Criticality of natural resources. *Journal of Environmental Management,* 2023. **325**.
8. Liu, L., X. Wang, and Z. Wang, Recent progress and emerging strategies for carbon peak and carbon neutrality in China. *Greenhouse Gases: Science and Technology,* 2023. **13**(5): p. 732–759.
9. Lane, J.E., Globalization and politics: Promises and dangers, in *Globalization and Politics: Promises and Dangers.* 2017: Taylor and Francis. 1–253.
10. Habiba, U. and C. Xinbang, The contribution of different aspects of financial development to renewable energy consumption in E7 countries: The transition to a sustainable future. *Renewable Energy,* 2023. **203**: p. 703–714.
11. Polewsky, M., *et al.*, Degrowth vs. green growth. A computational review and interdisciplinary research agenda. *Ecological Economics,* 2024. **217**.
12. Ha, Y.H. and J. Byrne, The rise and fall of green growth: Korea's energy sector experiment and its lessons for sustainable energy policy. *Wiley Interdisciplinary Reviews: Energy and Environment,* 2019. **8**(4).
13. Glavič, P., Evolution and current challenges of sustainable consumption and production. *Sustainability (Switzerland),* 2021. **13**(16).
14. Strzałkowski, A., Adaptation and operationalisation of sustainable degrowth for policy: Why we need to translate research papers into legislative drafts? *Ecological Economics,* 2024. **220**.

15. Erdiaw-Kwasie, M.O., Circularity challenges in SDGs implementation: A review in context, in *Sustainable Development Goals Series*, Michael Odei Erdiaw-Kwasie, and G. M. Monirul Alam, editors. 2023, Springer. p. 3–18.
16. Belmonte-Ureña, L.J., *et al.*, Circular economy, degrowth and green growth as pathways for research on sustainable development goals: A global analysis and future agenda. *Ecological Economics*, 2021. **185**.
17. Kallis, G., *et al.*, Research on degrowth. Annual *Review of Environment and Resources*, 2018. **43**: p. 291–316.
18. Trinh, V.L. and C.K. Chung, Renewable energy for SDG-7 and sustainable electrical production, integration, industrial application, and globalization: Review. *Cleaner Engineering and Technology*, 2023. **15**.
19. D'Adamo, I. and C. Sassanelli, A mini-review of biomethane valorization: Managerial and policy implications for a circular resource. *Waste Management and Research*, 2022. **40**(12): p. 1745–1756.
20. Deutz, P., H. Baxter, and D. Gibbs, Chapter 15: Governing resource flows in a circular economy: Rerouting materials in an established policy landscape, in *RSC Green Chemistry*, L.E. Macaskie, D.J. Sapsford, and W.M. Mayes, editors. 2020: Royal Society of Chemistry. p. 375–394.
21. Purnell, P., A.P.M. Velenturf, and R. Marshall, Chapter 16: New governance for circular economy: Policy, regulation and market contexts for resource recovery from waste, in *RSC Green Chemistry*, L.E. Macaskie, D.J. Sapsford, and W.M. Mayes, editors. 2020: Royal Society of Chemistry. p. 395–422.
22. Keohane, R.O. and D.G. Victor, The transnational politics of energy. *Daedalus*, 2013. **142**(1): p. 97–109.
23. Ralph, N. and L. Hancock, Energy security, transnational politics, and renewable electricity exports in Australia and South east Asia. *Energy Research and Social Science*, 2019. **49**: p. 233–240.
24. Amorós Molina, Á., *et al.*, Integrating the United Nations sustainable development goals into higher education globally: a scoping review. *Global Health Action*, 2023. **16**(1).
25. Suni, T., *et al.*, National future earth platforms as boundary organizations contributing to solutions-oriented global change research. *Current Opinion in Environmental Sustainability*, 2016. **23**: p. 63–68.
26. Zinchenko, V.V., *et al.*, Transformations of global cooperation processes as a strategy of sustainable social development in the context of research and education. in *E3S Web of Conferences*. 2021: EDP Sciences.
27. Akrofi, M.M., M. Okitasari, and R. Kandpal, Recent trends on the linkages between energy, SDGs and the Paris agreement: A review of policy-based studies. *Discover Sustainability*, 2022. **3**(1).
28. Hurrell, A., Geopolitics and global economic governance. *Oxford Review of Economic Policy*, 2024. **40**(2): p. 220–233.
29. Alam, A., *et al.*, Climate change and geopolitical conflicts: The role of ESG readiness. *Journal of Environmental Management*, 2024. **353**.
30. Feng, Y., *et al.*, Do the grey clouds of geopolitical risk and political globalization exacerbate environmental degradation? Evidence from resource-rich countries. *Resources Policy*, 2024. **89**.
31. Müller, F., Energy colonialism. *Journal of Political Ecology*, 2024. **31**(1).
32. Sachs, J., The new geopolitics. *Scientific American*, 2006. **294**(6): p. 30.

33. Ng, A. and J. Nathwani, Financial and technological innovation for sustainability: environmental, social and governance performance, in *Financial and Technological Innovation for Sustainability: Environmental, Social and Governance Performance*. 2023: Taylor and Francis. 1–258.

34. Kayachev, G., *et al.*, Expanding of green and renewable energy as a condition for economy transition to sustainable development, in *E3S Web of Conferences*. 2021: EDP Sciences.

35. Shu, Y., *et al.*, Geo-political risks, uncertainty, financial development, renewable energy, and carbon intensity: Empirical evidence from countries at high geo-political risks. *Applied Energy*, 2024. **376**.

36. Pata, U.K., S. Karlilar, and M.T. Kartal, On the road to sustainable development: The role of ICT and R&D investments in renewable and nuclear energy on energy transition in Germany. *Clean Technologies and Environmental Policy*, 2023. **26**, 2323–2335.

37. Vakulchuk, R., I. Overland, and D. Scholten, Renewable energy and geopolitics: A review. *Renewable and Sustainable Energy Reviews*, 2020. **122**.

38. Bourgoin, J., *et al.*, Mining resources, the inconvenient truth of the "ecological" transition. *World Development Perspectives*, 2024. **35**.

39. Yang, X. and L. Long, Renewable energy transition and its implication on natural resource management for green and sustainable economic recovery. *Resources Policy*, 2024. **89**.

40. Fischhendler, I., The role of borderlands in the energy transition: Toward a theoretical framework. *Environmental Innovation and Societal Transitions*, 2024. **53**.

41. Yasmeen, R. and W.U.H. Shah, Energy uncertainty, geopolitical conflict, and militarization matters for renewable and non-renewable energy development: Perspectives from G7 economies. *Energy*, 2024. **306**.

42. Wang, Q., C. Zhang, and R. Li, Impact of different geopolitical factors on the energy transition: The role of geopolitical threats, geopolitical acts, and geopolitical risks. *Journal of Environmental Management*, 2024. **352**.

43. Stergiou, A., Energy geopolitics revisited: Green economy instead of conflict, in *Contributions to International Relations*. 2022, Springer Nature. p. 85–95.

44. Zhang, D. and Q. Kong, Green energy transition and sustainable development of energy firms: An assessment of renewable energy policy. *Energy Economics*, 2022. **111**.

45. Tiwari, S. and K. Si Mohammed, Unraveling the impacts of linear economy, circular economy, green energy and green patents on environmental sustainability: Empirical evidence from OECD countries. *Gondwana Research*, 2024. **135**: p. 75–88.

46. Shah, S.Z.A., S. Chughtai, and B. Simonetti, Renewable energy, institutional stability, environment and economic growth nexus of D-8 countries. *Energy Strategy Reviews*, 2020. **29**.

47. Alsagr, N. and I. Ozturk, Natural resources rent and green investment: Does institutional quality matter? *Resources Policy*, 2024. **90**.

48. Steckel, J.C., *et al.*, From climate finance toward sustainable development finance. *Wiley Interdisciplinary Reviews: Climate Change*, 2017. 8(1).

49. Qadri, H.M.U.D., *et al.*, Mapping the evolution of green finance research and development in emerging green economies. *Resources Policy*, 2024. **91**.

50. Zhang, J., *et al.*, Spatiotemporal evolution of coordinated development between economic resilience and green finance under the background of sustainable development. *Sustainability (Switzerland)*, 2023. **15**(11).
51. Cao, Z. and L. Tao, Green finance and economic resilience: Investigating the nexus with natural resources through econometric analysis. *Economic Analysis and Policy*, 2023. **80**: p. 929–940.
52. Zeng, Q., C. Li, and C. Magazzino, Impact of green energy production for sustainable economic growth and green economic recovery. *Heliyon*, 2024. **10**(17).
53. Oxford researchers: Investing in green energy 'could save trillions of dollars'. *Fuel Cells Bulletin*, 2022. **2022**(10).
54. Chen, S.C.I., X. Xu, and C.M. Own, the impact of green finance and technological innovation on corporate environmental performance: Driving sustainable energy transitions. *Energies*, 2024. **17**(23).
55. Chen, J.M., M. Umair, and J. Hu, Green finance and renewable energy growth in developing nations: A GMM analysis. *Heliyon*, 2024. **10**(13).

Building Green Sustainable Energy Network

The universe of energy also provides choices that can help to address climate change. Green Sustainable Energy (GSE) provide low and zero carbon sources which will reduce Carbon dioxide (CO_2), the greenhouse gas responsible for most of the human-caused climate change in our atmosphere. Incisive decarbonization of the energy system through the use of GSE is needed to move toward a carbon-free world. GSE provides energy sources and a systems approach to achieving carbon neutrality. Green strategies will include the synergistic contributions of green finance, green technological innovation, and green energy adoption.

The need for energy in modern economic development has grown. Environmental pollution from conventional energy sources and their remaining limited amount are necessitating adoption of new energies. The search for sources of green sustainable energy must start with requirements, which will guide the selection of GSE sources. The GSE sources can be found in five "spheres" of the Earth ecosystem: "heliosphere" (sun), "lithosphere" (land to center of the Earth), "hydrosphere" (water), "biosphere" (living things), and "atmosphere" (air). Energy harvesting from these spheres, transformation, transport, and end use have significant impacts on the earth's environment, and environmental costs.

The energy must be captured and stored. Energy harnessing, harvesting, and storage accomplish these functions, in order fill the enormous power requirements of our world. Energy harnessing is a process that captures abundant energy in nature, while harvesting captures ambient energy. In these processes light, water, and wind, as well as motion, temperature gradients, electromagnetic radiation, and chemical energy can all become sources for green sustainable energy (GSE). The harnessed and harvested energies are converted into electrical energy. This energy is locally used directly, added to the energy grid, or stored in capacitors, batteries, or supercapacitors.

DOI: 10.1201/9781003407447-8

Chapter 6

Carbon neutrality, net-zero, and GSE

> *The climate crisis is about human security, economic security, environmental security, national security, and the very life of the planet ... It's more urgent than ever that we double down on our climate commitments.*
>
> Joe Biden, US President, COP27: UN summit, Nov 11, 2022

6.1 INTRODUCTION

Climate change is already affecting the entire world, with extreme and diverse weather conditions, such as drought, heat waves, heavy rain, and floods becoming more frequent. Other consequences of the rapidly changing climate include rising sea levels, ocean acidification, and loss of biodiversity. To limit global warming to 1.5 degrees Celsius – a threshold the Intergovernmental Panel for Climate Change (IPCC) suggests is safe – carbon neutrality by the mid-21st century is essential [1] (see Figure 6.1). The world must commit to a massive transition toward a carbon-neutral system.

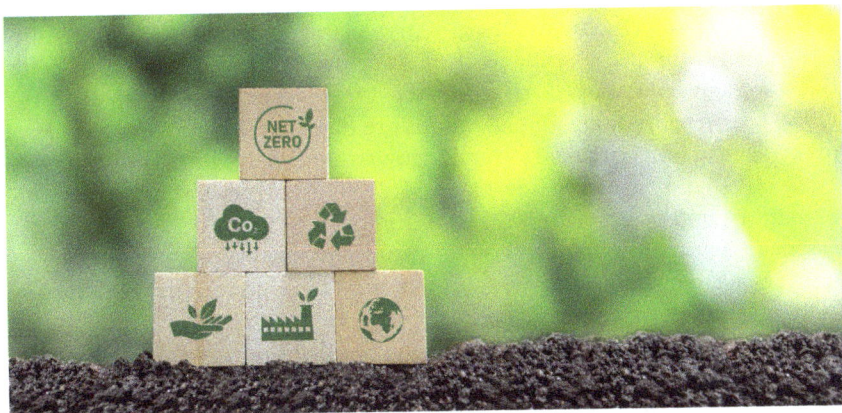

Figure 6.1 Carbon neutral and net zero concept for the natural environment.

Source: Shutterstock Photo ID: 2139611737.

DOI: 10.1201/9781003407447-9

This transition must include transforming the energy system from the current fossil fuel-based activities to carbon-free energy sources (see "Epilogue: The role of GSE in energy transition"). Although progress toward this goal varies across the globe in terms of intensity and depth, the overall trend of reducing human carbon footprint and striving toward a carbon-free energy system is evident.

Carbon neutrality requires striking a balance between releasing carbon to the atmosphere and absorbing carbon from the atmosphere. This carbon is in the form of carbon dioxide (CO_2). Removing CO_2 from the atmosphere may be accomplished in several ways [2]. Net zero is the ultimate goal and refers to the complete balance between the amount of greenhouse gas (GHG) produced and the amount that's removed from the atmosphere. Green sustainable energy (GSE) aids in this effort by adding more zero carbon sources. To accomplish net zero emissions, all worldwide greenhouse gas (GHG) emissions will have to be counterbalanced by carbon sequestration. The main natural carbon sinks are soil, forests, and oceans.

It is an empirical fact that climate warming is real, but its driving forces may not be known. Climate warming occurs with great uncertainty in the magnitude of the temperature increase; both human activities and natural forces contribute to climate change, but their relative contributions are difficult to quantify; the dominant role of the increase in the atmospheric concentration of greenhouse gases (including CO_2) in the current climate change is questioned by the scientific communities because of large uncertainties in the mechanisms of natural factors and anthropogenic activities and in the sources of the increased atmospheric CO_2 concentration [3]. Nevertheless, changes in societal activities, as occurred during Covid, do reduce CO_2 and lower average temperatures [4].

As a result of the growing global climate crisis, many countries have pledged to cut carbon dioxide emissions and other greenhouse gas emissions to achieve net-zero emission goals. These goals may be successfully realized with the use of environmental regulations, the adoption of green technology innovations, and a greater use of green sustainable energy (GSE). Green technological innovations are improved and facilitate times of financial development and economic growth. They have been shown to help reduce CO_2 emissions [5] (see Figure 6.2). This suggests that appropriate regulations and policies are necessary to attain net-zero carbon emissions. It is clear that there is a bidirectional causal link between green technology innovations, renewable energy, environmental taxes, economic growth, and CO_2 emissions.

6.2 CARBON NEUTRALITY AND NET-ZERO

Following on the preceding section, the increasing carbon emissions exert a devastating toll on human well-being, critical infrastructure, and natural

Figure 6.2 Reducing CO_2.

Source: Shutterstock Photo ID: 2497480211.

ecosystems, leaving a severe and costly path of destruction. The report of COP29 declared it necessary for OECD nations to initiate ambitious, unified strategies to address increasing CO_2 levels. Driven by ubiquitous solar and wind as main energy sources and batteries for storing electric power, the transition to green energy on the surface appears to be moving in the right direction. However, according to many published reports, the world is not on track to meet net-zero targets [6]. Inclusive green growth (IGG) captures the desired direction. IGG is a paradigm that harmonizes economic development with environmental sustainability and social equity, offering a clear path forward. By addressing the complex, interrelated nature of sustainability challenges, green policies (including GSE) offer a framework that combines diverse perspectives and expertise, driving transformative and profound change. Green sustainable energy can serve as a pathway for promoting environmental quality and reducing carbon emissions [7, 8].

CO_2, one of several greenhouse gases (GHG), significantly impacts climate change, adversely affecting both the environment and human well-being. As energy production and use are among the primary contributors to these emissions, it is wise to implement strategies to transition to renewable energy and reduce emissions by 2050–2070. Effective policies and pathways can achieve carbon neutrality, especially in the growing importance of hydrogen and biomass-generated energy. Key areas of energy use include alternative fuels, hydrogen, electric vehicles, solar and wind energy, carbon neutrality, and GHG mitigation. Factors such as population growth, urbanization rates, and coal consumption seem to have the opposite effect when compared to green sustainable energy adoption. GSE is emerging as a critical driver of environmental sustainability and achievement of net-zero emission goals. Policymakers are strongly encouraged to prioritize and implement optimal strategies that capitalize on opportunities to advance carbon neutrality objectives [9].

6.2.1 Chemistry of CO_2 formation and global warming

Carbon dioxide (CO_2) (see Figure 6.2) is the greenhouse gas responsible for most of the human-caused climate change in our atmosphere. It has the highest concentration in the atmosphere of any of the greenhouse gases. CO_2 is a direct product of both combustion and cellular respiration, causing it to be produced in great quantities both naturally and anthropogenically. Any time biomass or fossil fuels are burned, CO_2 is released. Major anthropogenic sources include electricity production from coal-fired and natural gas power plants, transportation, and industry. By 1990, a quantity of over seven billion tons of carbon (equivalent to 26 billion tons of carbon dioxide when the weight of the oxygen atoms is also considered) was being emitted into the atmosphere every year, much of it from industrialized nations. The action of the naturally existing greenhouse gases and anthropogenic greenhouse gases leads to an increase in the surface temperature of the Earth [10, 11].

When sunlight reaches Earth, the surface absorbs some of the light's energy and reradiates it as infrared waves (IR), which we feel as heat. These IR waves travel up into the atmosphere and will escape back into space if unimpeded. The presence of CO_2 and other greenhouse gases block the release of IR energy. Carbon dioxide, for example, absorbs energy at a variety of wavelengths between 2,000 and 15,000 nanometers – a range that overlaps with that of IR energy. As CO_2 soaks up this infrared energy, it vibrates and re-emits the infrared energy back in all directions. About half of that energy goes out into space, and about half of it returns to Earth as heat, contributing to the "greenhouse effect" [12]. Increased atmospheric CO_2 concentration therefore is widely being considered the main driving factor that causes the phenomenon of global warming. It is not totally clear that this is the only impact that CO_2 has on global warming. It should be stressed that the understanding of the functioning of Earth's complex climate system (especially for water, solar radiation, and so forth) is still poor, and, hence, scientific knowledge is not at a level to give definite and irrefutable answers for the causes of global warming [13].

Growth in fundamental drivers – energy use, economic productivity, and population – can provide quantitative indications of the definitive effect of increased CO_2 levels. Human energy expenditure in the Anthropocene has exceeded that used during the prior 11,700 years of human civilization, largely through combustion of fossil fuels. The global warming effect during the modern era is more than an order of magnitude greater still. Global human population, their productivity and energy consumption, and most changes impacting the global environment are highly correlated. This extraordinary outburst of consumption and productivity has forced clear physical, chemical, and biological changes to the Earth's stratigraphic record. This has included a significant increase in the global temperature (our new epoch has been named the Anthropocene) [14].

6.2.2 Environmental effect of increasing CO_2

The presence of increasing levels of CO_2 has an effect on the overall climate. Ecosystems and human civilization depend on predictable weather patterns, and these are changing rapidly due to the increase in GHGs in the atmosphere. The current average increase of 1.1 °C already has had noticeable effects on weather patterns and increased the risk of local flooding, droughts, wildfires, hurricanes, and damage to the ecosystem [15]. Remarkably, the current increase in global temperature has occurred in under 150 years and the ecosystem has very limited time to adapt to these changes. This is problematic and induces stress in the environmental system [16]. Slowing this increase in the earth's surface temperature as much as possible helps avoid catastrophic climate change.

In response to global climate change many countries have set a goal of "carbon neutrality" as a national strategy and put forward a vision of a zero-carbon future [17]. Environmental regulation (ER) and renewable energy technology innovation (RET) should be important actions that will contribute to achieving the carbon neutrality goal. It can be seen that there are direct and indirect effects of ER and RET on environmental sustainability from both theoretical and empirical dimensions. The results show that ER is beneficial to the reduction of CO_2 emissions when the level of economic development significantly frees money to address issues. RET has a beneficial impact on the environment, but when the economic development is low the carbon reduction effect of RET is reduced. Thus, the significance and magnitude of the carbon reduction effect of ER and RET depend on the regional economic development level (green sustainable energy will also have an effect on RET) [18]. This demonstrates a clear relationship between the reduction of CO_2, environmental programs, and overall environmental health.

6.2.3 Carbon neutrality strategies

Carbon neutrality means net-zero change in environmental carbon, which means that any actions that lead to emissions would be accompanied by other actions that confidently reduce – or offset – emissions [19]. The energy sector should focus on the integration of GSE sources, energy efficiency, and carbon capture technologies to provide the science and engineering to meet this goal. Promising developments in energy-efficiency methods and circular economy principles are strategies that are practiced in many industrial sectors. It is also important to include lifecycle assessment, economic complexity, and investments in shaping effective carbon neutrality strategies [20].

Carbon neutrality strategies play a significant role in climate mitigation efforts. Achieving carbon neutrality is a paramount objective for reducing greenhouse gas emissions and limiting the adverse impacts of anthropogenic activities on the environment. Carbon neutrality principles and technologies are important in climate mitigation. Importantly, there are a wide range of strategies possible across different sectors and industries that are effective and scalable [21]. Carbon neutrality solutions can achieve zero atmospheric carbon dioxide emissions by targeting greenhouse gas emissions in various sectors [22, 23]. These strategies include renewable energy transitions, energy efficiency improvements, policy and regulatory frameworks, carbon pricing mechanisms, nature-based solutions like afforestation and reforestation, advanced technologies like carbon capture, hydrogen economy, and circular economy practices. Coordinating these solutions, recognizing trade-offs, and aligning them for a low-carbon future is crucial.

The transition to GSE is a key component in the move toward carbon neutrality. Carbon neutrality must include a transition from reliance on fossil fuels to GSE sources like wind, solar, and hydroelectric power, as these are zero carbon. Governments should encourage this transition through financial mechanisms like subsidies and tax incentives and by setting renewable energy targets and mandates [24, 25].

One of the major challenges in this transition is the intermittent variability of some of these energy sources [26], which necessitates the development of advanced energy storage solutions and grid modernization to manage supply and demand effectively. (These will be discussed in subsequent chapters.) The move to GSE also depends on the availability of rare earth minerals, which are limited in quantity [27]. As described in this book's last chapter, phasing out fossil fuel infrastructure presents economic and social challenges, including job losses, capital investment, and ensuring a just transition for affected workers and communities.

6.3 GREEN SUSTAINABILITY, CARBON NEUTRALITY, AND THE CIRCULAR ECONOMY

Two United Nations Sustainable Development Goals help to measure the impact of sustainability efforts in the world. These goals are: SDG 7 Affordable and Clean Energy, and SDG 13 Climate Action. Both SDG 7

Figure 6.3 Carbon neutral and green sustainability.

Source: Shutterstock Photo ID: 2139818309.

and SDG 13 set goals and reveal weaknesses in how energy and climate issues are being addressed [28]. New policies and priorities are needed to be implemented. The failure to improve – that is to reduce energy consumption and move toward net zero – demonstrates the overall lack of progress in decoupling economic growth from energy consumption (see Figure 6.3). Three key policy recommendations that will aid the process can be made: (1) "creative *carbon* accounting" cannot be a means to success in SDGs; (2) economic growth must be decoupled from energy consumption; and (3) national and regional economic recovery plans must prioritize energy efficiency improvements (including for the poorest households), and GSE sources need to be framed in terms of resiliency for the environment and society [28].

To achieve a limit of the global temperature increase to 1.5 °C, the current energy system needs to change. Currently, about 80% of our primary energy supply is still being met with fossil energy [29]. These fossil fuels are coal, oil, and gas. Extracting these from the ground and combusting them adds carbon to our atmosphere and increases the global average temperature. Incisive decarbonization must limit the use of these fuels. In order to meet the goals set out in the Paris Agreement and move toward a circular economy and a carbon-free world, carbon neutrality is critical, and green sustainability will help.

6.3.1 Green sustainability

Green sustainability refers to the responsible management of natural resources to fulfill current needs without compromising the ability of future generations to meet their natural resource needs. This sustainability aims to balance ecological, economic, and social goals, such as reducing carbon emissions, promoting renewable energy, and ensuring equitable resource access. The practice of green sustainability requires the resources and tools that GSE provides.

Green energy that is sustainable is created with the least amount of adverse environmental effects possible. In this way, GSE preserves and prolongs sources for future generations. The continued use of fossil fuels emits greenhouse gases, which are the main contributor to global warming and climate change, thus negatively affecting the environment. GSE sources, including hydro, geothermal, solar, and wind energy, contribute little or nothing to the global warming process. Carbon emissions are global problems, which must be solved by global efforts. Thus carbon neutrality must be a global concern.

6.3.2 Carbon neutrality

The goal of carbon neutrality is to protect the planet for future generations. The climate crisis driven largely by human-induced carbon emissions threatens ecosystems, biodiversity, and the stability of our global climate. By committing to carbon neutrality, individuals, companies, and even entire countries aim to contribute to the global mission of minimizing global warming. Every ton of CO_2 reduced or offset helps slow down climate change's impact, from rising sea levels to more extreme weather patterns. The significance of carbon neutrality goes beyond environmental benefits. This will also benefit public health – especially in cities where high emission levels often lead to pollution and poor air quality. Carbon neutrality helps drive positive social impacts, as cleaner air means fewer respiratory issues and healthier communities.

To achieve carbon neutrality, the development and application of low-carbon, zero-carbon, and negative-carbon technologies are crucial in energy and industrial activities [26]. Carbon neutrality involves carefully measuring emissions across all activities – from energy use and transportation to production processes – and reducing these emissions as much as possible. While remaining emissions may be offset through initiatives like tree planting, investing in renewable energy, or supporting carbon capture technologies, it is important to stop their generation at the source. Carbon neutrality functions as a framework to guide companies and individuals in reducing their climate impact by focusing on low/no carbon sources and aligning efforts toward global net-zero goals.

It's also important to distinguish carbon neutrality from a "carbon zero" activity, which would require no CO_2 emissions at all – an extremely challenging, if not impossible, target for most industries. "Net zero" often describes a more comprehensive, long-term approach aimed at reducing emissions as close to zero as possible and this is before offsetting the residual emissions that cannot be eliminated. As will be discussed in the next chapter on GSE sources, "carbon zero" is one of the characteristics of GSE sources.

6.3.3 Carbon emissions reduction through a circular economy

Despite an immense amount of work on global warming and climate change mitigation technologies, work is now focusing on the changing role of CO_2 in industrial processes, which relates to the introduction of circular economy (CE) principles. CO_2 is regarded as a valuable resource in CE. This primarily results from carbon sequestration projects: carbon capture and storage; carbon capture, utilization, and storage; and carbon capture and utilization. CO_2 is a valuable and sought-after resource for various industries. Due to its increasing value, the use of sequestration technologies as a tool in CE will benefit industries [30]. Carbon neutrality benefits from this effort to capture carbon dioxide (CO_2) emitted into the atmosphere and reuse it in industrial processes. This capture and reuse of CO_2 can also provide a means of GSE storage. To make the goal of carbon neutrality attainable, the paradigm of CE will be helpful.

Tangible successes in achieving the SDGs have been limited, underscoring the critical need for innovative approaches to fostering energy performance and reducing *carbon* emissions. Adopting CE principles as a strategic pathway to mitigate environmental, social, and economic challenges and promote sustainable, net-zero-energy solutions is both desirable and viable. As discussed later in this chapter, this will require source development to work in tandem with green engineering. There is a synergy between CE and the SDGs, particularly in resource use circularity and GSE integration. The circular economy can bridge the gap between sustainability, renewable energy adoption, and climate change mitigation efforts [31]. Specifically, there are three pivotal ways GSE can work with CE to effect carbon neutrality:

- Help define how GSE sources fit in the CE paradigm;
- Qualify appropriate times to use/reuse GSE sources;
- Utilize GSE tools to facilitate CO_2 reduction.

A systematic methodology is needed to reverse the continued release of carbon emissions which is producing negative effects on human well-being, critical infrastructure, and natural ecosystems. Inclusive green growth, a paradigm that harmonizes economic development with environmental sustainability and social equity, offers a clear path forward. CE can integrate these elements into a cohesive response that will drive toward a sustainable future. The circular economy provides a unified strategy to achieve carbon neutrality. Integrated policies within CE significantly enhance the effectiveness of green growth and energy transitions, ensuring equitable benefits across all societal segments. By addressing the complex, interrelated nature of sustainability challenges, these policies offer a robust framework consolidating diverse perspectives and expertise, driving transformative and profound change [7].

6.3.4 Green strategies for carbon neutrality

There is a pivotal role for green strategies in achieving carbon neutrality through the synergistic contributions of green finance, green technological innovation, and green energy adoption. Studies have shown the significance of green finance mechanisms in mobilizing resources for sustainable initiatives, including renewable energy projects and energy-efficient technologies. There is a catalytic effect from green technological innovation in driving technological advancements, reducing emissions, and fostering economic growth. This leads to the transformative potential of clean energy adoption, which can reduce carbon footprints and bolster the transition to a low-carbon economy. The critical nexus of green strategies and carbon neutrality offers a roadmap for a more sustainable and environmentally responsible future [32].

At the heart of green growth and green economy are technological innovations that can result in greening the production processes and products. Innovation in green technologies can be enhanced by policy instruments, universities as knowledge-producing institutions acting as engines of economic growth, and cross-sector university partnerships. Employing increased awareness of energy efficiency and renewable energy technology is essential; this is a role for GSE. The existing strategy for achieving a carbon-neutral economy is insufficient and it is now essential to explore additional techniques to enhance the current strategy. Currently, there is an inverse relationship between green growth and carbon neutrality, and this relationship must change. The relationship is dynamic, indicating a significant adoption and promotion of green growth for the pursuit of carbon neutrality and sustainable long-term economic growth is desirable. The pressure from environmental health and economic growth must be balanced. Incorporating environmentally related innovations and technologies by using the circular economy will result in enhanced energy efficiency [33].

6.4 GSE CONTRIBUTION TO CARBON NEUTRALITY

The availability of natural resources for energy generation is seemingly unlimited. Global energy strategies must prioritize achieving a sustainable equilibrium by harnessing their development and use [34]. It is imperative to safeguard the security of our shared future, which must include GSE. Although today's environmental concerns might seem focused on high energy production and ever-increasing transportation, failing to prudently manage our resources is not only unwise, it is immoral.

As we have demonstrated in this chapter, a significant challenge today is adopting clear strategies and programs concerning decreasing carbon dioxide (CO_2) emissions. Attempting to accomplish carbon neutrality will be challenging during times of energy crisis, price volatility, and war [35]. Green sustainable energy is produced with the least amount of adverse environmental effects possible and it provides a supportive aid to reducing CO_2 (see Figure 6.4). GSE sources, including hydro, geothermal, solar, and wind energy, are alternatives to the use of fossil fuels and contribute little or nothing to the global warming process [36].

Figure 6.4 GSE generation on the Earth.

Source: Shutterstock Photo ID: 2547021705.

6.4.1 Energy variables and key points

GSE combines positive environmental and sustainability qualities, but at present, these energy sources can often involve steep initial costs for setup and deployment. Fossil fuel energies usually come at a lower price and are more readily available but bear significant ecological consequences [37]. The allure of GSE lies in their sustainability, environmental benefits, and progressively competitive economic positioning. However, a massive shift to these energy frameworks will demand considerable financial outlays and an overhaul of the extant infrastructure. (See the final chapter "Epilogue: The role of GSE in energy transition.")

The following tables will provide a good overview of the potential GSE sources and technologies.

In Table 6.1 we lay out the criteria that new GSE sources must address in order to meet the prevalent energy demands in the 21st century. These criteria will affect their utility and desirability.

The factors in Table 6.1 establish parameters and constraints on GSE sources. This will allow us to provide in Table 6.2 an initial overview of the GSE technologies and resources that will be discussed in greater detail in the next chapter. As changes occur due to advancements in technology or

Table 6.1 Energy variables and key points [38]

GSE variables	Key points
Energy content	The energy content of a particular resource should be quantified in Joules or British thermal units and be sufficient to meet existing requirements.
Power density	The energy achievable from a specific source, often expressed as watts per square meter. (This factor helps in assessing the feasibility of various energy resources.)
Renewability	The capacity of a resource to regenerate naturally over time (contrary to those which will eventually be depleted.)
Energy return on investment (EROI)	The EROI indicates the economic feasibility of a resource and the environmental impact of its extraction and use.
Environmental impact	The environmental consequences arising from harvesting and using a resource encompass elements such as CO_2 emissions, pollution, and disruptions to natural habitats.
Cost	The monetary costs tied to the extraction, processing, and distribution of the resource. This becomes the economic price of a resource compared to other sources.
Technological maturity	This indicates the current and projected sophistication of the technologies and methods used for resource extraction and application.
Geopolitical context	Describes the interplay of political, economic, and social factors that can impact the availability, cost, and dependability of energy.
Scalability	The capacity of the resource to scale up in response to growing energy needs.

Table 6.2 Overview of GSE technological resources and their characteristics [38]

Renewal technologies	Characteristics
Solar power technologies	These involve direct conversion of sunlight into electrical energy. Solar technologies find diverse applications in power generation, water heating, and even vehicle propulsion.
Wind power technologies	These involve wind turbines, which transform the motion energy of the wind into electricity (onshore or offshore). Available in a range of sizes (small individual units to commercial-scale wind farms).
Hydroelectric power technologies	Usually hydroelectric dams, which produce electricity by capturing the energy of cascading water, and micro hydro systems (electricity for individual residences or small communities).
Geothermal power technologies	Geothermal power facilities, which produce electricity by utilizing the Earth's crust heat, and geothermal heating systems.
Bioenergy technologies	Technologies that produce biofuels like ethanol and biodiesel, as well as systems designed for generating electricity and heat from biomass.
Nuclear power technologies	Nuclear power facilities that generate electricity by utilizing the energy produced through nuclear reactions.

alterations in demand patterns and shifts in geopolitical factors, the desirability of the individual sources will shift [39].

6.5 GREEN SUSTAINABLE ENERGY PROMOTING ZERO-CARBON

The current predominant energy systems are reliant upon the supply chain and technical capabilities of fossil fuels [40]. This is a well-developed energy system that balances supply and demand via a very liquid market. Both oil and coal are produced in various regions across the globe and transported virtually everywhere. There is also a massive system of pipelines transporting natural gas to industrial sites and consumers, as well as a growing liquefied natural gas system. The current transport and trading system, in combination with the high energy density of fossil fuels, is well established and firmly entrenched in global economies.

In order to compete with or even replace this established chain adopting GSE must cause a paradigm shift. GSE has significant potential to drive a clean energy transition. GSE sources are being developed as low- or non-carbon alternatives to conventional energy sources. The main goal of developing green energy technologies is to provide energy in a sustainable manner while cutting down on waste and greenhouse gas emissions, thus reducing the overall carbon footprint of energy production [36]. The development and use of GSE sources can significantly contribute to the progress of slowing climate change by offering "zero-carbon" solutions (see Figure 6.5).

Figure 6.5 Carbon net zero.

Source: Shutterstock Photo ID: 2195336915.

6.5.1 Foundational science

Foundational science and technology will be required to facilitate the transformative changes necessary to reduce CO_2 and avoid catastrophic climate-induced change. Accomplishing this will require massive upscaling of research that can enhance these transformations, especially in the realm of GSE [41]. The science and technology, if it is truly to lead to a societal transformation, must be characterized by the following:

(1) Drive transformations to low-carbon, resilient living
(2) Develop and utilize practical knowledge
(3) Research both the energy sources and system applications
(4) Transcend current thinking
(5) Take a multi-faceted approach to understand and shape change
(6) Acknowledge the value of contributions from diverse disciplines.

6.5.2 Specific areas that benefit from GSE

About 80% of the global population lives in countries that are net-importers of fossil fuels [42]. This means that approximately six billion people are dependent on fossil fuels coming from other countries, which makes them vulnerable to geopolitical shocks and crises. Green sustainable energy sources are available in all countries, and their potential is yet to be fully harnessed. These energy sources offer ways for countries to diversify their economies and protect themselves from the unpredictable price swings of fossil fuels while driving inclusive economic growth, new jobs, and poverty alleviation. GSE will both benefit and contribute to the following areas germane to carbon neutrality efforts:

6.5.2.1 GSE source development

Harnessing sources that are renewable and zero-carbon must be the hall-mark of GSE. For instance, solar photovoltaic (PV), wind energy, and other green sustainable energies provide carbon-free renewable energy to reach ambitious global carbon-neutrality goals, but their yields are in turn influenced by future climate change. The co-benefits in enhancing and stabilizing renewable energy sources demonstrate beneficial reciprocity in achieving global carbon neutrality and highlight the increasing demand for these resources in future decades [43]. Exploratory work in this area and other fields is necessary to uncover new sources of energy.

6.5.2.2 Biomass and biofuel production

Under what circumstances can biomass and biofuels be considered GSE? Carbon-neutral liquid fuels, in the form of biofuels and fuels synthesized using renewable energy, provide a route to mitigating the climate change and energy security concerns, which currently challenge the transport sec-tor without the paradigm shifts in vehicle technology and infrastructure required by electrification of the vehicle fleet or conversion to a hydrogen economy. The low-carbon-number alcohols, biodiesel, and kerosene offer the prospects of continued high levels of affordable mobility through the gradual evolution of the vehicle fleet and fuel distribution infrastructure to one that is broadly compatible with that which pervades today [44].

6.5.2.3 Green engineering

Increasing GSE in light of carbon neutrality will involve green innovation, geopolitical risk, and increased capital. There is a pivotal role for green technology innovation and green engineering in building the systems and structures that will deliver these energies. The combined influence of green technology and green engineering are critical in lowering CO_2. Green engineering will provide necessary technologies when transitioning toward cleaner and GSE sources. Economic growth, when coupled with green technological advancements, can pave the way for environmental sustainability [45].

6.5.2.4 Green politics and artificial intelligence

The rapid advancement of artificial intelligence (AI) in the 21st century is driving profound societal changes and playing a crucial role in optimizing energy systems to achieve carbon neutrality. Many nations have developed national AI strategies and are advancing AI applications in energy, manufacturing, and agriculture sectors to meet this goal. However, there are disparities among nations, that need to be addressed for regulatory consistency and fair distribution of AI benefits. At this point geopolitical risk can hinder or enhance efforts to attain carbon neutrality through energy transition. From 1990 to 2022, carbon neutrality has improved primarily due to technological advancements. Energy transition accelerates carbon neutrality in both developed and developing countries. Strategies that improve AI and uphold geopolitical stability are crucial to achieve carbon neutrality [46].

6.5.2.5 Green finance

Carbon neutrality is an effective way to deal with issues such as global warming and extreme climate disasters. Green finance promotes the green transformation of economy through a series of financial instruments and plays an important role in achieving the goal of carbon neutrality. The development of green finance is conducive to the green transformation of production activities, reducing energy consumption and improving energy efficiency, as well as driving green innovation and stimulating green economic growth [47].

6.5.2.6 Sustainable development

Carbon neutrality demonstrates an active response to climate change and recognizes communities with a shared vision for humanity's future. In the medium and long term, the carbon neutrality goal will lead to global economic and social development changes, forcing economic growth to shift from high-input, low-efficiency, and high-pollution to low-input, high-efficiency, low-pollution, and high-quality development. These are characteristics of sustainable development. According to the projections of different institutions, in the next three decades, many developing countries will promote the utilization of renewable resources, energy efficiency improvement, and increased electrification. New energy vehicles and home furnishing continue to promote wind power, photovoltaics, energy storage, hydrogen energy, and smart grids. Sustainable development will depend on strategic emerging industries, high-tech industries, and green environmental protection industries to be new drivers of economic growth engines of the green economy [48].

6.5.2.7 Supply chain and redefining distribution networks

The transition to GSE will de-centralize energy distribution. The new carbon-neutral supply chains are more agile and resilient and help firms to capitalize on the opportunities for reducing overall supply chain costs. GSE can move toward carbon neutrality by effectively developing supply chains focused on expanding GSE adoption and reducing carbon emissions. Mapping these supply chains and looking into supply chain processes, activities, and entities in real time help to reduce carbon emissions [49].

This will evolve into the reconfiguration of distribution networks. Switching to green sustainable energy is complicated by the traditional large central generation and distribution system. Most of the new energy systems, such as solar, wind, geothermal, hydropower, and biogas, are smaller-scale distributed systems. Consumers can also be producers with their own solar panels or even turbines, capitalizing on periods during sunlight or the high wind. Flexible and large-scale storage is also needed in this case. So, apart from a shift away from fossil fuel systems to renewable and low-carbon systems, the distribution and storage network itself needs to transition as well to a complex multiple generation and distribution system (see "Epilogue: The role of GSE in energy transition"). We need a much more flexible system with a higher degree of intelligence and responsiveness, which will require increased energy storage capacity [50].

6.6 "RIGHT THING TO DO"

Carbon neutrality/zero-carbon is an ambitious goal that has been put forward as achievable on or before 2060. Initially, most of the current energy policies focus more on carbon emission reduction, efficiency, and high penetration of renewable energy. As technology improves, it will be important to revise the current energy policies to incorporate a carbon neutrality framework moving to zero carbon [51].

Since there is a scientific consensus that the energy sector is a heavy factor leading the planet to the tipping point of climate change, transitioning to sustainable energy sources is not only scientifically realistic to halt foreseeable climatic adversities, but it is also the proper course of action to take. The empirical results show that through lowering CO_2 intensity and greenhouse gas emissions, the Earth's health returns. Green sustainable energy transition helped to lower greenhouse gas and CO_2 intensity. These results demonstrate that environmental policies supporting commitment to achieving low-carbon development goals are beneficial. To secure industrial emissions reduction for a future with net-zero carbon emissions, using policies that reduce environmental emissions, such as the carbon regulations, will be beneficial. Additionally, plans for the sustainable energy transition that include a quick rise in renewable energy sources in the overall energy mix are successful in lowering environmental emissions. For environmental sustainability and low-carbon development, it is thus advised to divert the taxation burden from renewable energy technologies to the fossil fuel industry to enhance the sustainable energy transition phenomenon for achieving Sustainable Development Goals (especially SDG 7 and SDG 13) [52].

Social innovation, circularity, and GSE transition may all be considered vital environmental, social, and governance (ESG) components to promote sustainability. To further support the use of government aid in promoting GSE in carbon neutrality, current and anticipated advancements are needed in these areas. Implementing ESG practices will benefit greatly from GSE. Future directions for both firms and governments should pursue macro-level goals concerning energy transition and circularity through social innovation [53].

REFERENCES

1. Stille, L., M. Wilson, and K. Bishop, Introduction to the energy transition, in *Geophysics and the Energy Transition*, Malcolm Wilson, Tom Davis, and Martin Landro, editors. 2024, Elsevier. 3–13.
2. Béres, R., M. Junginger, and M.V.D. Broek, Assessing the feasibility of CO2 removal strategies in achieving climate-neutral power systems: Insights from biomass, CO2 capture, and direct air capture in Europe. *Advances in Applied Energy*, 2024. **14**.
3. Fang, J.Y., *et al.*, Global warming, human-induced carbon emissions, and their uncertainties. *Science China Earth Sciences*, 2011. **54**(10): p. 1458–1468.
4. Amir, M. and S.Z. Khan, Assessment of renewable energy: Status, challenges, COVID-19 impacts, opportunities, and sustainable energy solutions in Africa. *Energy and Built Environment*, 2022. **3**(3): p. 348–362.
5. Sakilu, O.B. and H. Chen, Realizing carbon neutrality in top-emitter countries: Do green technology innovation, renewable energy, financial development, and environmental tax matters? *Sustainability (Switzerland)*, 2025. **17**(1).
6. Singh, R., Expediting green energy transition [Expert view]. *IEEE Power Electronics Magazine*, 2024. **11**(2): p. 74–77.
7. Shobande, O.A., L. Ogbeifun, and A.K. Tiwari, Carbon neutrality: Synergy for energy transition, circular economy and inclusive green growth. *Journal of Environmental Management*, 2025. **374**.
8. Shobande, O.A., L. Ogbeifun, and A.K. Tiwari, Re-evaluating the impacts of green innovations and renewable energy on carbon neutrality: Does social inclusiveness really matters? *Journal of Environmental Management*, 2023. **336**.
9. Soni, R., A. Dvivedi, and P. Kumar, Carbon neutrality in transportation: In the context of renewable sources. *International Journal of Sustainable Transportation*, 2025. **19**: p. 1–15.
10. Curran, J.C. and S.A. Curran, Natural sequestration of carbon dioxide is in decline: climate change will accelerate. *Weather*, 2025. **80**: 85–87.
11. Akther, S., *et al.*, Exploring the influence of green growth and energy sources on "carbon-dioxide emissions": implications for climate change mitigation. *Frontiers in Environmental Science*, 2024. **12**.

12. Fecht, S. *How Exactly Does Carbon Dioxide Cause Global Warming?* February 25, 2021.

13. Florides, G.A. and P. Christodoulides, Global warming and carbon dioxide through sciences. *Environment International*, 2009. 35(2): p. 390–401.

14. Syvitski, J., *et al.*, Extraordinary human energy consumption and resultant geological impacts beginning around 1950 CE initiated the proposed Anthropocene Epoch. *Communications Earth and Environment*, 2020. 1(1).

15. Diffenbaugh, N.S., *et al.*, Quantifying the influence of global warming on unprecedented extreme climate events. *Proceedings of the National Academy of Sciences of the United States of America*, 2017. 114(19): p. 4881–4886.

16. Shah, I.H., *et al.*, Comprehensive review: Effects of climate change and greenhouse gases emission relevance to environmental stress on horticultural crops and management. *Journal of Environmental Management*, 2024. 351.

17. Pender, K., *et al.*, Future strategies for decarbonisation of carbon fibre products: A roadmap to net zero 2050. *Journal of Cleaner Production*, 2025. 486.

18. Hao, Y., X. Li, and M. Murshed, Role of environmental regulation and renewable energy technology innovation in carbon neutrality: A sustainable investigation from China. *Energy Strategy Reviews*, 2023. 48.

19. Kopalle, P.K., *et al.*, Delivering affordable clean energy to consumers. *Journal of the Academy of Marketing Science*, 2024. 52(5): p. 1452–1474.

20. Al Khaffaf, I., A. Tamimi, and V. Ahmed, Pathways to carbon neutrality: A review of strategies and technologies across sectors. *Energies*, 2024. 17(23).

21. Evro, S., B.A. Oni, and O.S. Tomomewo, Global strategies for a low-carbon future: Lessons from the US, China, and EU's pursuit of carbon neutrality. *Journal of Cleaner Production*, 2024. 461.

22. Zhang, L., J. Ling, and M. Lin, Carbon neutrality: a comprehensive bibliometric analysis. *Environmental Science and Pollution Research*, 2023. 30(16): p. 45498–45514.

23. Ho, Y.S. and M.H. Wang, Comments on "Carbon neutrality: A comprehensive bibliometric analysis". *Environmental Science and Pollution Research*, 2024. 31(52): p. 61969–61970.

24. Awosusi, A.A., *et al.*, Can green resource productivity, renewable energy, and economic globalization drive the pursuit of carbon neutrality in the top energy transition economies? *International Journal of Sustainable Development and World Ecology*, 2023. 30(7): p. 745–759.

25. Gasparatos, A., *et al.*, Renewable energy and biodiversity: Implications for transitioning to a Green Economy. *Renewable and Sustainable Energy Reviews*, 2017. 70: p. 161–184.

26. Paraschiv, L.S. and S. Paraschiv, Contribution of renewable energy (hydro, wind, solar and biomass) to decarbonization and transformation of the electricity generation sector for sustainable development. *Energy Reports*, 2023. 9: p. 535–544.

27. Jowitt, S.M., Renewable energy and associated technologies and the scarcity of metal, in *Living With Climate Change*. 2024, Elsevier. 45–63.

28. LaBelle, M.C. and T. Szép, Green Economy: Energy, Environment, and Sustainability, in *Contributions to Economics*, László Mátyás, editor. 2022, Springer Science and Business Media Deutschland GmbH. 325–364.

29. Gentile, G. and J. Gupta, Orchestrating the narrative: The role of fossil fuel companies in delaying the energy transition. *Renewable and Sustainable Energy Reviews*, 2025. **212**.

30. Tcvetkov, P., A. Cherepovitsyn, and S. Fedoseev, The changing role of CO2 in the transition to a circular economy: Review of carbon sequestration projects. *Sustainability (Switzerland)*, 2019. **11**(20).

31. Cervantes Puma, G.C., A. Salles, and L. Bragança, Nexus between urban circular economies and sustainable development goals: A systematic literature review. *Sustainability (Switzerland)*, 2024. **16**(6).

32. Qamruzzaman, M. and S. Karim, Unveiling the synergy: Green finance, technological innovation, green energy, and carbon neutrality. *PLoS ONE*, 2024. **19**(10).

33. Mamman, S.O. and K. Sohag, Green growth: A strategy for carbon neutrality, in *Recent Developments in Green Finance, Green Growth and Carbon Neutrality*, Muhammad Shahbaz, Kangyin Dong, Daniel Balsalobre-Lorente, and Ayfer Gedikli, editors. 2023, Elsevier. 301–319.

34. Hussain, A., S.M. Arif, and M. Aslam, Emerging renewable and sustainable energy technologies: State of the art. *Renewable and Sustainable Energy Reviews*, 2017. **71**: p. 12–28.

35. Halkos, G.E. and P.S.C. Aslanidis, Green energy pathways towards carbon neutrality. *Environmental and Resource Economics*, 2024. **87**(6): p. 1473–1496.

36. Sharma, V.K., *et al.*, A comprehensive review of green energy technologies: Towards sustainable clean energy transition and global net-zero carbon emissions. *Processes*, 2025. **13**(1).

37. Singh, A. and P. Baredar, Techno-economic assessment of a solar PV, fuel cell, and biomass gasifier hybrid energy system. *Energy Reports*, 2016. **2**: p. 254–260.

38. Soni, N., *et al.*, Advancing sustainable energy: Exploring new frontiers and opportunities in the green transition. *Advanced Sustainable Systems*, 2024.8: 2400160 (1 of 23).

39. Sinsel, S.R., R.L. Riemke, and V.H. Hoffmann, Challenges and solution technologies for the integration of variable renewable energy sources: A review. *Renewable Energy*, 2020. **145**: p. 2271–2285.

40. English, B.C., R.J. Menard, and B. Wilson, The economic impact of a renewable biofuels/energy industry supply chain using the renewable energy economic analysis layers modeling system. *Frontiers in Energy Research*, 2022. **10**.

41. Fazey, I., *et al.*, Ten essentials for action-oriented and second order energy transitions, transformations and climate change research. *Energy Research and Social Science*, 2018. **40**: p. 54–70.

42. Osman, A.I., *et al.*, Cost, environmental impact, and resilience of renewable energy under a changing climate: A review. *Environmental Chemistry Letters*, 2023. **21**(2): p. 741–764.

43. Lei, Y., *et al.*, Co-benefits of carbon neutrality in enhancing and stabilizing solar and wind energy. *Nature Climate Change*, 2023. **13**(7): p. 693–700.

44. Pan, X., *et al.*, Toward efficient biofuel production: A review of online upgrading methods for biomass pyrolysis. *Energy and Fuels*, 2024. **38**(20): p. 19414–19441.

45. Zhao, Y., *et al.*, Racing towards zero carbon: Unraveling the interplay between natural resource rents, green innovation, geopolitical risk and environmental pollution in BRICS countries. *Resources Policy*, 2024. **88**.
46. Salman, M., *et al.*, G20 roadmap for carbon neutrality: The role of Paris agreement, artificial intelligence, and energy transition in changing geopolitical landscape. *Journal of Environmental Management*, 2024. **367**.
47. Shahbaz, M., *et al.*, Recent developments in green finance, green growth and carbon neutrality, in *Recent Developments in Green Finance, Green Growth and Carbon Neutrality,* Muhammad Shahbaz, Kangyin Dong, Daniel Balsalobre-Lorente, and Ayfer Gedikli, editors. 2023: Elsevier. 1–434.
48. Zhang, Y., *et al.*, Optimizing sustainable development in arid river basins: A multi-objective approach to balancing water, energy, economy, carbon and ecology nexus. *Environmental Science and Ecotechnology*, 2025. **23**.
49. Jiang, H., *et al.*, Role of supply chain digitalization and global supply chain in decarbonization of natural resources sector supply chain. *Journal of Environmental Management*, 2024. **370**.
50. Oduro, R.A. and P.G. Taylor, Future pathways for energy networks: A review of international experiences in high income countries. *Renewable and Sustainable Energy Reviews*, 2023. **171**.
51. Alabi, T.M., *et al.*, Strategic potential of multi-energy system towards carbon neutrality: A forward-looking overview. *Energy and Built Environment*, 2023. **4**(6): p. 689–708.
52. Jabeen, G., *et al.*, Promoting green taxation and sustainable energy transition for low-carbon development. *Geoscience Frontiers*, 2025. **16**(1).
53. Popescu, C., *et al.*, Social innovation, circularity and energy transition for environmental, social and governance (ESG) practices: A comprehensive review. *Energies*, 2022. **15**(23).

Chapter 7

Sources of green sustainable energy

Most people think of solar and wind as new energy sources. In fact, they are two of our oldest.

Michael Shellenberger [1]

7.1 INTRODUCTION

The need for energy in modern economic development has grown. This is being heightened by the elevated level of environmental pollution from conventional energy sources and their remaining limited amounts. Globalization is leading the modern world toward green and sustainable sources of energy (see Figure 7.1). Green energy draws on clean sources that have a lower environmental impact compared to conventional energies. The energy systems of the different regions and territories vary based on various parameters, so a detailed green energy planning for sustainable development is desired to procure a suitable common strategy for energy management [2]. This chapter will begin the process by providing information on potential sources of green sustainable energy (GSE) and the energy management strategies that will aid the researchers and decision-makers in applying the procedures.

Figure 7.1 Sources of green sustainable energy.

Source: Shutterstock Photo ID: 2261372101.

DOI: 10.1201/9781003407447-10

The global energy landscape is gradually being transformed. GSE reduces greenhouse gas emissions, diversifies the energy supply, and lowers dependence on volatile and uncertain fossil fuel markets. In response, there is growing support for green energy sources, a factor that could help accelerate the current energy transition. Despite these positive developments, much remains to be done globally to make the energy transition. The demand for oil, at least in the short and medium term, will increase, due to a considerable projected increase in global energy demand, which cannot be met only through the development of renewable energy sources. However, it is better for countries to start pursuing a policy of energy diversification to become less dependent on trends in the energy market and have more reliable sources of energy [4].

A necessary first step is the recognition of viable green energy sources. GSE sources will provide energy that will not increase climate change and will play an important role in improving energy security and accessibility. The efforts of every country to strengthen the energy sector through the development of green energies will reduce geopolitical risks and external costs for society. The large-scale use of these energies will contribute to sustainable development [3].

The energy and climate crises present challenges and opportunities for scientists to find solutions in the field of green sustainable energy sources. For example, the search for new opportunities in the energy industry revealed the potential use of hydrogen as an energy source. To fully realize its use as an energy source, barriers remain for designing safe, usable, reliable, and effective forms of hydrogen storage. This highlights one of the issues facing new sources of GSE. The emergence of new sources necessitates new ways of storage. New ways of energy storage are vital in a GSE-based economy [4].

7.2 SEARCH FOR GSE SOURCES

The search for sources of green sustainable energy must start with requirements. The standards for green sustainable energy are guidelines for supplying energy from GSE sources. Policies should ensure energy is produced from eligible energy resources (see Figure 7.2). These guidelines apply to all energy supplied. Although no international GSE regulations exist, there is growing consensus for its development. GSE standards will cover sources like:

- Solar
- Geothermal
- Wind
- Biomass
- Hydrogen

Figure 7.2 Criteria of sustainable green energy.

Source: Shutterstock Vector ID: 523872340.

A stable energy supply is possible with the successful use of GSE and this provides energy security. The rising need for energy, caused by both population growth and economic activity, makes this imperative. Nations must find a way to meet energy demands while also making sure it is inexpensive and sustainable. The transition to GSE must include new energy resources, as they could lessen reliance on fossil fuels, decrease environmental consequences, and be sustainable. Indeed, GSE is plentiful, clean, and might one day provide all energy needs. This means that investing in GSE energy sources is becoming the preferred option [5].

7.2.1 Existential need

Oil and gas will account for 44% of the world's energy supply in 2050, compared to the current 53%. Gas will become the largest single source of energy by the end of the first third of the 21st century. The fossil fuel share of the world's primary energy mix will reduce from 81% currently to 52% in 2050. Gas will continue to play a key role alongside renewables in helping to meet future, lower carbon, energy requirements [6]. These sources won't last forever, nor are they most desirable. The current statistics reveal changes are occurring.

This is a critical time for climate action, rapid and far-reaching transitions depend on the choices and actions that are taken now and in the near term. The inevitable depletion of fossil fuel sources should provide new opportunities. Integrated approaches that advance the discovery of sources for sustainable development of energy and environment systems will play crucial roles in enabling this shift in direction and will better protect the life-support systems of the planet. Green sustainable energy systems open new paths for carbon neutrality by employing green hydrogen and possibilities for wind and wave energy. New advances reveal the efficient energy production from biowaste and wastewater treatment. Boosting solar energy utilization is supported by advances in solar photovoltaic planning and photothermal conversion performance [7].

7.2.2 Criteria for GSE sources

The supply of global energy by the mid-21st century will mostly depend on nontraditional energy sources. Substantial technology investments in the development of renewable energy sources (RESs) over the past 25–30 years have reduced the cost of energy by orders of magnitude. Some RESs approached traditional hydrocarbon energy both in performance and in their economic indicators [8]. This will be more important as the transition involves a new GSE. An economic evaluation of prospects for GSE development now becomes both a scientific and practical problem. Current work will help measure the effects of GSE sources on global and domestic energy and determine the speed at which the transition will occur.

Providing sustainable energy from GSE systems will face many challenges. The identification and development of GSE sources will need to address concerns in four areas (known as the four drivers of green innovation). These are the scientific driver, the economic driver, the environmental driver, and the political driver. No one driver is predominant, but all will influence, and at times dictate, the possibilities available. Accomplishing the energy transition to GSE will require a multifaceted approach and all of the drivers will be important. The following examples and sections highlight critical areas of focus and will correspond to the drivers that dictate the GSE choices:

- Investment in green sustainable energy infrastructures(economic)
- Energy efficiency measures (environmental)
- Technology innovation and research and development (scientific)
- Policy support, regulatory frameworks, and global cooperation (political)

7.2.2.1 Scientific drivers

Green sustainability in science must address problems afflicting the environment, devise measures for improving economies and societies, and ultimately enhance the lives of people. Scientific evidence is the basis for achieving sustainability in areas relating to our common future and green sustainable energies. This work will involve various sciences, including mathematics, biology, agriculture, computer science, engineering, and physics [9].

7.2.2.2 Economic drivers

Economic growth and the development of global markets are coupled with energy use, which has caused an increase in global energy demand and created pressure on the supply of energy resources. It is critical to understand the importance of multidisciplinary economic approaches that integrate social and technological perspectives to solve current sustainability problems. Economics will both promote the development of sustainable energy and sustainable production and will affect its distribution and use [10].

7.2.2.3 Environmental drivers

There are challenges and opportunities for green sustainable energy and environment transition when facing the dilemma of meeting growing energy demand while reducing greenhouse gas emissions and environmental vulnerability. Policies are needed that can achieve economic growth, social welfare, and environmental sustainability efficiently and effectively. Green energy and environmental policies, such as developing a clear and consistent policy framework, enhancing regional cooperation and collaboration, leveraging information technology and data analytics, emphasizing sustainability and resilience, and engaging with other stakeholders and partners are invaluable [11].

7.2.2.4 Political drivers

In a complex and intricately connected world, it is imperative to engage and partner with policymakers to articulate an energy policy that is not only scientifically and technically sound but also one that the global society will accept. The energy challenge is the type of problem that we, as a global society, have never faced before and we need decisive scientific and political leadership to address it [12].

7.3 GREEN SUSTAINABLE ENERGY SOURCES

The use of GSE sources will save energy costs and protect the environment in the long term (see Figure 7.3) A wise methodology is to map selected

Figure 7.3 Sources of sustainable green energy.

Source: Shutterstock Photo ID: 2201000683.

energy and environmental compartments and evaluate GSE sources, using the four drivers. At present, energy supply is still mostly provided by sources coming from fossil or nuclear reserves. There is an "upper limit of exhaustibility" of fossil resources, but disagreements exist between experts in the field of the life expectancy of the reserves [10]. Given this situation, the first step involves identifying replacement sources.

There are increasing new areas in the development of alternative energy sources. If GSE sources are appropriate, then GSE power generation must become comparable to traditional sources of generation. To achieve this there must be an increase in the use of alternative energy sources and a decrease in the use of hydrocarbon fuels [9]. The potential areas from which to draw GSE are vast in number, and the sources can be identified by partitioning the environment into spheres.

It is held that everything in Earth's ecosystem can be placed into one of four major subsystems: land (including all layers within the Earth), water, living things, or air [13]. In addition to these four subsystems, a fifth subsystem can be added: the Sun. Together these may be called spheres: the "heliosphere" (sun), "lithosphere" (land to the center of the Earth), "hydrosphere" (water), "biosphere" (living things), and "atmosphere" (air). In the sections that follow we will show how each of these five spheres contributes potential GSE sources.

7.3.1 Heliosphere

The heliosphere is the magnetosphere, astrosphere, and outermost atmospheric layer of the Sun. It takes the shape of a vast, tailed bubble-like region of space. The "bubble" of the heliosphere is continuously "inflated" by plasma originating from the Sun, known as the solar wind. This active heliosphere is constantly providing energy to the planets in the solar system [14]. The Sun releases an enormous amount of energy during explosive solar activities, such as solar flares and coronal mass ejections. The solar corona can be heated up to tens of millions of degrees and a large number of charged particles can be accelerated to nearly the speed of light. Heated plasmas and high-energy particles increase solar radiation across the entire

Table 7.1 Examples of methods to harness energy from the heliosphere

Method	Description	Points or issues	Reference
Solar voltaic systems (PV)	PV system or solar power system supplies usable solar power by means of photovoltaics (use of solar panels)	Is mismatched with the frequency of the modern power system	[10, 11, 17]
Heat capture	solar thermal collector collects heat by absorbing sunlight	High initial cost and requires a large footprint	[18, 19]
Photosynthesis	Using light to drive natural or synthetic reactions	Available sunlight or light of specific wavelengths	[20, 21]

electromagnetic spectrum, from radio to gamma-ray wavelengths, which can have a profound effect on the Earth's upper atmosphere. These create additional ionization and heating in the Earth's upper atmosphere [15].

The sun provides a nearly unlimited supply of energy. We have not fully tapped into either its variety or its supply. Energy from the Sun makes it possible for life to exist on Earth. It is responsible for photosynthesis in plants, vision in animals, and many other natural processes, such as the movements of air and water that create weather. Infrared radiation from the Sun is responsible for heating the Earth's atmosphere and surface. Without energy from the Sun, Earth would freeze. Harvesting sustainable energy from the sun and cold space to uninterruptedly generate green electricity is a potential alternative to address energy needs. The recent advances in energy harvesters (solar absorbers or/and radiative coolers) integrated with energy converters (thermoelectric generators) provide uninterrupted energy harvest and electricity generation, but challenges remain [16].

Table 7.1 provides examples of green and sustainable methods to harness energy from the heliosphere.

7.3.2 Atmosphere

The earth-atmosphere balances incoming energy from the Sun and outgoing energy from the Earth. The energy released from the Sun is emitted as shortwave light and ultraviolet energy. The earth-atmosphere energy balance is achieved as the energy received from the Sun balances the energy lost by the Earth back into space. The absorption of infrared radiation trying to escape from the Earth back to space is particularly important to the global energy balance. Energy absorption by the atmosphere stores more energy near its surface than it would if there was no atmosphere [22].

Atmospheric electricity describes the electrical charges in the Earth's atmosphere (or that of another planet). The movement of charge between

Table 7.2 Examples of methods to harness energy from the atmosphere

Method	Description	Points or issues	References
Hydrogen	Hydrogen is an energy vector that is generated by solar energy breaking up H_2O in the atmosphere.	Must overcome challenges of producing green hydrogen (technological intricacies, economic barriers, societal considerations)	[10, 24, 25]
Green ammonia	Ammonia, produced in the atmosphere, has easy storage at low pressure and ambient or low temperature and ambient pressure.	Technological intricacies, economic barriers, societal considerations, and far-reaching policy implications are major hurdles.	[26–28]
Green methanol	Sustainable methanol is produced using renewable electricity, sustainable hydrogen (H_2), and recycled carbon dioxide (CO_2).	Currently costly	[29–31]
Energy from atmospheric water	Hydrovoltaic technology, especially moisture-induced electricity, shows great potential in harvesting energy from atmospheric water.	Limited operational time and requires advanced system designs	[32]
Wind capture	Wind turbines collect and convert the kinetic energy that wind produces into electricity to add power to the grid.	Implementation has environmental impacts (environmental impact of wind energy systems, particularly on-shore and off-shore wind turbines)	[33–35]

the Earth's surface, the atmosphere, and the ionosphere is known as the global atmospheric electrical circuit. Atmospheric electricity involves electrostatics, atmospheric physics, meteorology, and Earth science [23].

Table 7.2 provides examples of green and sustainable methods to harness energy from the atmosphere.

7.3.3 Hydrosphere

The hydrosphere is the total amount of water on Earth. The hydrosphere includes water that is on the surface of the planet, underground, and in the air. Earth's hydrosphere can be liquid, vapor, or ice. On Earth, liquid water exists on the surface in the form of oceans, lakes, and rivers. It also exists below ground – as groundwater, in wells, and aquifers. Water vapor is most visible as clouds and fog.

Table 7.3 Methods to harness energy from the hydrosphere

Method	Description	Points or issues	References
Wave power	Ocean wave power generation techniques exist (converting wave energy into electrical energy).	Design, control, efficiency, and safety of ocean wave power generation systems are challenging.	[37, 38]
Green hydrogen	Generation of hydrogen through a chemical process (electrolysis).	Scalability of green hydrogen production faces challenges including infrastructure gaps, energy losses, excessive power consumption, and high costs throughout the value chain.	[25, 39, 40]
Salinity gradient ("Osmotic power")	The salinity gradient energy ("blue energy") widely exists at the junction of river water and seawater.	The main waste product of salinity gradient technology is brackish water.	[41, 42]

Water moves through the hydrosphere in a cycle. Water collects in clouds, then falls to Earth in the form of rain or snow. This water collects in rivers, lakes, and oceans. Then it evaporates into the atmosphere to start the cycle all over again. There are tremendous amounts of energy in this movement.

The energy associated with all the water movement has the potential to provide vast amounts of power worldwide. It has yet to be properly harnessed, but tidal power could eventually become one of our main – and cleanest – sources of GSE. Due to high cost and lack of research, hydrospheric energy has not been fully exploited yet. The technology used in harnessing marine energy presents a threat to aquatic life, but suitable methods and focused research in this area can lead to harnessing this abundant source of renewable energy [36].

Table 7.3 provides examples of green and sustainable methods to harness energy from the hydrosphere.

7.3.4 Lithosphere to the Earth's core

The lithosphere is the solid, outer part of Earth. The lithosphere usually refers to the brittle upper portion of the mantle and the crust, but it includes the outermost layers of Earth's structure. It is bounded by the atmosphere above and the asthenosphere (another part of the upper mantle) below, thereby exchanging and transferring energy. In the context of GSE sources we will also include the mass from the actual crust to the liquid core of the Earth since it provides more energy.

Table 7.4 Examples of methods to harness energy from the lithosphere and internal energy

Method	Description	Points or issues	References
Geothermal heat	Geothermal energy is heat energy from the Earth. Resources are reservoirs of hot water at varying temperatures and depths below the Earth's surface.	Scaling up of these technologies has technical and non-technical challenges.	[43–45]
Gravitational energy	Gravitational energy or gravitational potential energy is the potential energy an object with mass has due to the gravitational potential of its position in a gravitational field.	Using the force of gravity to generate electricity through the fall of water from a height is common, but technologies are needed.	[46–48]
Energy geostructures	Energy geostructures are innovative technologies that combine the functions of structural support and geothermal energy harvesting.	Not trivial and more research is needed.	[49–51]

Table 7.4 provides examples of green and sustainable methods to harness energy from the lithosphere and Earth's internal mass.

7.3.5 Biosphere

The biosphere, also called the ecosphere, is the worldwide sum of all ecosystems. It can also be termed the zone of life on Earth. The biosphere (which is technically a spherical shell) is virtually a closed system regarding matter, with minimal inputs and outputs. Regarding energy, it is an open system, with photosynthesis capturing solar energy. By the most general biophysiological definition, the biosphere is the global ecological system integrating all living beings and their relationships, including their interaction with the elements of the lithosphere, cryosphere, hydrosphere, and atmosphere [52, 53].

Table 7.5 provides examples of green and sustainable methods to harness energy from the biosphere.

7.3.6 Searching for sources

Whether it is the efficient conversion of solar radiation into stable, energy-dense liquid energy carriers, harnessing hydrogen, tapping geological sources, or discovering novel energies, the process of looking for alternative

Table 7.5 Examples of methods to harness energy from the biosphere

Method	Description	Points or issues	References
Waste to Energy	Waste-to-energy (WtE) is a source for future electricity generation from Municipal Solid Waste (MSW).	Advanced technologies in developing countries are still in the infancy level with a few exceptions.	[54–56]
Biohydrogen	Biohydrogen is a very promising and environmentally feasible fuel alternative.	Sustainable green methods of conversion of biomass are needed.	[57–59]
Biocrude	Fuels from a wide range of biomass and/or waste materials.	Hydrothermal liquefaction (HTL) and other technologies are needed.	[11, 60–62]
Biomass and agro-waste	Biomass energy comes from various trees and other plants like perennial grasses, waste and landfill gases.	Cost of production and biomass characterization are major issues.	[63, 64]
Microalgae	Biofuel production with low cost, high speed of growth, and ability to grow in harsh environments.	Photosynthetic processes consume CO_2 and solar light to grow and provide a promising bioenergy source	[8, 9]

energy sources is driven by the increasing demand for energy. This must be guided by environmental lessons and sustainability imperatives. Sustainable approaches and technological advancements to produce green energy from the environment or organic sources will continue to supply alternatives to fossil fuels. The combination of novel technologies with the existing production lines will play a vital role in the generation of GSE [5].

7.4 GSE PRODUCTION

The world uses and produces many different types and sources of energy, which can be grouped into general categories such as primary, secondary, renewable, or fossil fuels. GSE mostly comes from primary sources and will be described in the following sections (see Figure 7.4). Electricity is a secondary energy source that is generated (produced) from primary energy sources. In this section we will look at the amounts of energy coming from different sources.

Energy production, transformation, transport, and end use have significant impacts on the Earth's environment, and environmental costs are

Figure 7.4 Sustainable green energy production.

Source: Shutterstock Photo ID: 2467829013.

usually associated with the emissions – whether they are of a thermal, chemical, nuclear, or other natural processes. One means of reducing the environmental impact of emissions is to increase the efficiency of resource utilization.

One of the major challenges of designing technologies to reduce their environmental impact involves determining an environmentally optimal configuration or selecting the most appropriate one from competing options. This determination or selection is made difficult by the complex relationship between the technology being considered and the characteristics of the residuals produced. Process or technology changes generally affect the flow rate and composition of all resulting effluent streams [65, 66].

7.4.1 Renewable energy sources

Renewable energy sources are essential to green sustainable energy, as they generally strengthen energy security and emit far fewer greenhouse gases than fossil fuels. Renewable energy projects sometimes raise significant sustainability concerns, such as risks to biodiversity when areas of high ecological value are converted to bioenergy production or wind or solar farms.

Hydropower is the largest source of renewable electricity while solar and wind energy are growing rapidly. Photovoltaic solar and onshore wind are the cheapest forms of new power generation capacity in most countries. For more than half of the 770 million people who currently lack access to electricity, decentralized renewable energy such as solar-powered mini-grids is likely the cheapest method of providing it by 2030. United Nations targets for 2030 include substantially increasing the proportion of renewable energy in the world's energy supply [67].

According to the International Energy Agency, renewable energy sources like wind and solar power are now commonplace sources of electricity, making up 70% of all new investments made in the world's power generation [68]. The Agency expects renewables to become the primary energy source for electricity generation globally in the next three years, overtaking coal.

7.4.2 Solar

The Sun is Earth's primary source of energy, a clean and abundantly available resource in many regions. In 2019, solar power provided around 3% of global electricity, mostly through solar panels based on photovoltaic cells (PV). Solar PV is expected to be the electricity source with the largest installed capacity worldwide by 2027 [69]. The cost of electricity from new solar farms is competitive with, or in many places, cheaper than electricity from existing coal plants. Various projections of future energy use identify solar PV as one of the main sources of energy generation in sustainability [70].

Solar PV generation increased by a record 320 TWh (up 25%) in 2023, reaching over 1 600 TWh. It demonstrated the largest absolute generation growth of all renewable technologies in 2023. This generation growth rate is close to the level envisaged from 2023 to 2030 in the Net Zero Emissions by 2050 (NZE) Scenario [71].

7.4.3 Wind power

Wind has been an important driver of development over millennia, providing mechanical energy for industrial processes, water pumps, and sailing ships. Modern wind turbines are used to generate electricity and provide approximately 6% of global electricity in 2019. Electricity from onshore wind farms is often cheaper than existing coal plants and competitive with natural gas and nuclear. Wind turbines can also be placed offshore, where winds are steadier and stronger than on land but construction and maintenance costs are higher [72, 73].

Significantly growing wind energy is being contemplated as one of the main avenues to reduce carbon footprints and decrease global risks associated with climate change. However, obtaining a comprehensive perspective on wind energy considering the many diverse factors that impact its development and growth is challenging. Significant factors in the evolution of wind energy are technological advancement, financial viability, environmental concerns, government incentives, and the impact of wind on the ecosystem. Wind energy may continue growing all over the world as long as all the factors critical to its development are addressed [74].

7.4.4 Hydropower

Hydroelectric plants convert the energy of moving water into electricity. In 2020, hydropower supplied 17% of the world's electricity, down from a high of nearly 20% in the mid-to-late 20th century. Compared to reservoir-based facilities, river-flow hydroelectricity generally has less environmental impact. However, its ability to generate power depends on river flow, which can vary with daily and seasonal weather. Reservoirs provide water quantity controls that are used for flood control and flexible electricity output while also providing security during drought for drinking water supply and irrigation [75].

Hydropower ranks among the energy sources with the lowest levels of greenhouse gas emissions per unit of energy produced, but levels of emissions vary enormously between projects. The highest emissions tend to occur with large dams in tropical regions. It is apparent that these facilities do not produce carbon dioxide; however, their significant negative impacts on the environment are still found and cannot be ignored [76].

7.4.5 Geothermal

In 2022, 24 countries, including the United States, generated about 92 billion kWh of electricity from geothermal energy. Indonesia was the top geothermal electricity producer at about 17 billion kWh – which was about 5% of Indonesia's total electricity generation. Kenya was the seventh-highest geothermal electricity producer, at about 5 billion kWh, which was equal to about 45% of Kenya's annual electricity generation. Kenya had the largest percentage share of electricity generation from geothermal energy among all countries with geothermal power plants [77–79].

7.4.6 Bioenergy

The climate impact of bioenergy varies considerably depending on where biomass feedstocks come from and how they are grown. For example, burning wood for energy releases carbon dioxide; those emissions can be significantly offset if the trees that were harvested are replaced by new trees in a well-managed forest, as the new trees will absorb carbon dioxide from the air as they grow. However, the establishment and cultivation of bioenergy crops can displace natural ecosystems, degrade soils, and consume water resources and synthetic fertilizers [80].

Integration of different renewable energies like solar, wind, and hydroenergy technologies in combination with bioenergy for hybrid systems maximizes energy output through a 15–37% decrease in total system cost and minimizes waste of resources. It also contributes to achieving climate goals and energy security accompanied by an eminent 12–30% decrease in greenhouse gas emissions. The advantages, limitations, and future directions of cutting-edge biomass conversion technologies and their integration with artificial intelligence and hybrid systems reduce waste and promote sustainable development, aligning with the circular economy framework and sustainability goals for a clean and green future [81].

7.4.7 Marine energy

Marine energy has the smallest share of the energy market. It includes OTEC, tidal power, which is approaching maturity, and wave power, which is earlier in its development. There is a global need for producing more clean energy from renewable sources. Significant electrical power can be extracted from marine tidal currents. However, the power harnessed from marine tidal currents highly fluctuates due to the swell effect and the periodicity of the tidal phenomenon. To improve the power quality and make the marine generation system more reliable, energy storage systems can play a crucial role [82, 83].

7.5 GREEN SUSTAINABLE ENERGY'S FUTURE

Considering the climate crisis and further forecasted ecological catastrophes, the world has begun to recast its view of energy (see Figure 7.5). This is shown by increased efforts to find new sources of energy. According to calculations by the International Energy Agency (IEA), in 2022, a record growth in the use of renewable energy sources (RES) occurred in the world – up to 320 GW of new capacities. The governments of most countries increasingly understand the growing role of RES in ensuring energy security [84]. Green innovation, technological innovations, foreign direct investment, and medium-high-tech exports enhance both short and long-run economic growth. These findings substantiate the desirability of the Sustainable Development Goals (SDGs) and furnish evidence that expands theoretical frameworks, emphasizing the significance of green innovation, technological innovation, and export competitiveness in economic development. This demands that countries require a holistic strategy to enhance technological innovation and export competitiveness, foster green infrastructure and advanced industries, address inefficiencies in the renewable energy sector, and implement regulatory measures to ensure sustainable economic growth and drive green innovation [85].

Examining an updated outlook of global energy production and utilization shows how distributed generation from GSE will provide the key

Figure 7.5 Energy's future.

Source: Shutterstock Photo ID: 1865740750.

solution to meeting global energy requirements. Policymakers need guide-lines for energy systems that will be instrumental in ensuring the transition from fossil fuels to combustion-free GSE for all energy uses [86].

Following the path of GSE and decarbonizing the energy system is a key approach to achieving energy balance. Therefore, identifying GSE sources, analyzing their environmental impacts, and developing their economics are essential to the Earth's survival [12].

REFERENCES

1. *Michael Shellenberger Quotes*. BrainyQuote.com. October 5, 2024. Available from: https://www.brainyquote.com/quotes/michael_shellenberger_1114355.
2. Fernandez, M.I., *et al.*, Review of challenges and key enablers in energy systems towards net zero target: Renewables, storage, buildings, & grid technologies. *Heliyon*, 2024. **10**(23).
3. Androniceanu, A. and O.M. Sabie, Overview of green energy as a real strategic option for sustainable development. *Energies*, 2022. **15**(22).
4. Jastrzębski, K. and P. Kula, Emerging technology for a green, sustainable energy promising materials for hydrogen storage, from nanotubes to graphene—a review. *Materials*, 2021. **14**(10).
5. As'ad, S., Why renewable energy gained attention and demand globally? *Nature Environment and Pollution Technology*, 2024. **23**(1): p. 467–473.
6. Green, J., Ed. World News: Natural gas to become primary energy source by 2035. *Pipeline and Gas Journal*, 2017. **244**(10).
7. Krajačić, G., *et al.*, Sustainable development of energy, water and environment systems in the critical decade for climate action. *Energy Conversion and Management*, 2023. **296**.
8. Porfiriev, B.N. and S.A. Roginko, Energy on renewable sources: Prospects for the world and for Russia. *Herald of the Russian Academy of Sciences*, 2016. **86**(6): p. 433–440.
9. Akinsemolu, A.A., The principles of green and sustainability science, in *The Principles of Green and Sustainability Science*. 2020: Springer Singapore. 1–387.
10. Sáez-Martínez, F.J., *et al.*, Drivers of sustainable cleaner production and sustainable energy options. *Journal of Cleaner Production*, 2016. **138**(Part 1): p. 1–7.
11. Qudrat-Ullah, H., A review and analysis of green energy and the environmental policies in South Asia. *Energies*, 2023. **16**(22).
12. Abruña, H.D., Energy in the age of sustainability. *Journal of Chemical Education*, 2013. **90**(11): p. 1411–1413.
13. Nelson, W.M., Sustainable agricultural chemistry in the 21st century: Green chemistry nexus, in *Sustainable Agricultural Chemistry in the 21st Century: Green Chemistry Nexus*. 2023: CRC Press. 1–294.
14. Müller, D., *et al.*, Solar orbiter: Exploring the sun-heliosphere connection. *Solar Physics*, 2013. **285**(1–2): p. 25–70.
15. Kong, X., J. Liu, and G. Li, Editorial: New insights into high-energy processes on the Sun and their geospace consequences. *Frontiers in Astronomy and Space Sciences*, 2025. **12**.

16. Zhang, S., *et al.*, An emerging energy technology: Self-uninterrupted electricity power harvesting from the sun and cold space. *Advanced Energy Materials*, 2023. **13**(19).
17. Rehman, T.U., *et al.*, Global perspectives on advancing photovoltaic system performance: A state-of-the-art review. *Renewable and Sustainable Energy Reviews*, 2025. **207**.
18. Jamali, H., Analysis of parabolic trough solar collector thermal efficiency with application of a graphene oxide nanosheet-based nanofluid. *Energy Science and Engineering*, 2024. **12**(7): p. 2974–2991.
19. Gaurav, K. and S.K. Verma, Critical review on thermal performance enhancement techniques for flat plate solar collectors. *Proceedings of the Institution of Mechanical Engineers, Part E: Journal of Process Mechanical Engineering*, 2024.
20. Romero, E., V.I. Novoderezhkin, and R. Van Grondelle, Quantum design of photosynthesis for bio-inspired solar-energy conversion. *Nature*, 2017. **543**(7645): p. 355–365.
21. Jiang, F., F.H. Kleiner, and M.E. Aubin-Tam, Harnessing photosynthesis for materials, devices, and environmental technologies. *Current Opinion in Biotechnology*, 2025. **92**.
22. Skirvin, S.J. and T. Van Doorsselaere, Mode conversion and energy flux absorption in the structured solar atmosphere. *Astronomy and Astrophysics*, 2024. **683**.
23. Harrison, R.G., *et al.*, Focus on high energy particles and atmospheric processes. *Environmental Research Letters*, 2015. **10**(10).
24. Kaheel, S., *et al.*, Advancing hydrogen: A closer look at implementation factors, current status and future potential. *Energies*, 2023. **16**(24).
25. Jayachandran, M., *et al.*, Challenges and opportunities in green hydrogen adoption for decarbonizing hard-to-abate industries: A comprehensive review. *IEEE Access*, 2024. **12**: p. 23363–23388.
26. Machaj, K., *et al.*, Ammonia as a potential marine fuel: A review. *Energy Strategy Reviews*, 2022. **44**.
27. Müller, M., *et al.*, Comparison of green ammonia and green hydrogen pathways in terms of energy efficiency. *Fuel*, 2024. **357**.
28. Dey, S., *et al.*, Green Ammonia: An alternative sustainable energy source for clean combustion, in *energy, Environment, and Sustainability*. 2024, Springer. p. 11–24.
29. Hank, C., *et al.*, Economics & carbon dioxide avoidance cost of methanol production based on renewable hydrogen and recycled carbon dioxide-power-to-methanol. *Sustainable Energy and Fuels*, 2018. **2**(6): p. 1244–1261.
30. Reddy, V.J., *et al.*, Sustainable E-fuels: Green hydrogen, methanol and ammonia for carbon-neutral transportation. *World Electric Vehicle Journal*, 2023. **14**(12).
31. Tabibian, S.S. and M. Sharifzadeh, Statistical and analytical investigation of methanol applications, production technologies, value-chain and economy with a special focus on renewable methanol. *Renewable and Sustainable Energy Reviews*, 2023. **179**.
32. Tan, J., *et al.*, Harvesting energy from atmospheric water: Grand challenges in continuous electricity generation. *Advanced Materials*, 2024. **36**(12).

33. Alam, M.S., *et al.*, Solar and wind energy integrated system frequency control: A critical review on recent developments. *Energies,* 2023. **16**(2).

34. García-Alonso, R., *et al.* Analysing barriers to achieving SDG 7. Managing green product development in the wind energy sector, in *IFIP Advances in Information and Communication Technology.* 2023. Springer Science and Business Media Deutschland GmbH.

35. Ha, K., *et al.*, Recent control technologies for floating offshore wind energy system: A review. *International Journal of Precision Engineering and Manufacturing - Green Technology,* 2021. **8**(1): p. 281–301.

36. Pokazeev, K.V., D.A. Solovyev, and L.V. Nefedova, New opportunities for the development of renewable sources of hydrosphere energy, in *Springer Proceedings in Earth and Environmental Sciences.* 2022, Springer Nature. p. 203–209.

37. Wang, J., Z. Chen, and F. Zhang, A review of the optimization design and control for ocean wave power generation systems. *Energies,* 2022. **15**(1).

38. Mwasilu, F. and J.W. Jung, Potential for power generation from ocean wave renewable energy source: A comprehensive review on state-of-the-art technology and future prospects. *IET Renewable Power Generation,* 2019. **13**(3): p. 363–375.

39. Kumari, P. and A. Kumar, Solar assisted generation of plasmonic silver photocatalyst for wastewater remediation and green hydrogen production. *Energy and Climate Change,* 2025. **6**.

40. Rahman, M.J. and M.R.S. Tuser. Green hydrogen revolution: A comprehensive analysis and future outlook for sustainable energy transition in Bangladesh, in *7th International Conference on Development in Renewable Energy Technology, ICDRET 2024.* 2024. Institute of Electrical and Electronics Engineers Inc.

41. Hsu, W.S., *et al.*, Miniaturized salinity gradient energy harvesting devices. *Molecules,* 2021. **26**(18).

42. Hussain, A., S.M. Arif, and M. Aslam, Emerging renewable and sustainable energy technologies: State of the art. *Renewable and Sustainable Energy Reviews,* 2017. **71**: p. 12–28.

43. Nkinyam, C.M., *et al.*, Exploring geothermal energy as a sustainable source of energy: A systemic review. *Unconventional Resources,* 2025. **6**.

44. Anya, B., *et al.*, Exploring geothermal energy based systems: Review from basics to smart systems. *Renewable and Sustainable Energy Reviews,* 2025. **210**.

45. Loschetter, A., *et al.*, Integrating geothermal energy and carbon capture and storage technologies: A review. *Renewable and Sustainable Energy Reviews,* 2025. **210**.

46. Lee, J., *et al.*, Harnessing gravitational, hydrodynamic and negative dielectrophoretic forces for higher throughput cell sorting. *Biochip Journal,* 2012. **6**(3): p. 229–239.

47. Aziz, M., Advanced green technologies toward future sustainable energy systems. *Indonesian Journal of Science and Technology,* 2019. **4**(1): p. 89–96.

48. Aziz, M.Q.A., J. Idris, and M.F. Abdullah, Simulation of the conical gravitational water vortex turbine (GWVT) design in producing optimum force for energy production. *Journal of Advanced Research in Fluid Mechanics and Thermal Sciences,* 2022. **89**(2): p. 77–91.

49. Ravera, E., A.F. Rotta Loria, and L. Laloui, Performance of complex energy geostructures. *Geomechanics for Energy and the Environment*, 2024. **38**.
50. Ouzzine, B., *et al.*, Scaling laws for the modelling of energy geostructures. *International Journal of Physical Modelling in Geotechnics*, 2024. 24(4): p. 202–216.
51. Kong, G.Q., *et al.*, Review on the evaluation of ground-coupled heat pump and energy geostructures to exploit shallow geothermal energy with regional scale. *Yantu Lixue/Rock and Soil Mechanics*, 2024. 45(5): p. 1265–1283.
52. Eriksson, K.E. and K.H. Robèrt, From the Big Bang to sustainable societies. *Acta Oncologica*, 1991. 30(6 Spec No): p. 5–14.
53. Chen, J., *et al.*, Does renewable energy matter to achieve sustainable development goals? The impact of renewable energy strategies on sustainable economic growth. *Frontiers in Energy Research*, 2022. **10**.
54. Roque, B.A.C., *et al.*, Hydrogen-powered future: Catalyzing energy transition, industry decarbonization and sustainable economic development: A review. *Gondwana Research*, 2025. **140**: p. 159–180.
55. Liang, F.Y., *et al.*, Waste-to-resource strategy through green synthesis of PET-derived metal-organic frameworks for efficient photocatalytic dye degradation. *Microporous and Mesoporous Materials*, 2025. **384**.
56. Kumar, A., *et al.*, From waste to fuel: Harnessing high specificity lipases from Candida rugosa fermentation for sustainable biodiesel. *Bioresource Technology Reports*, 2025. **29**.
57. Togun, H., *et al.*, Current developments in the use of nanotechnology to enhance the generation of sustainable bioenergy. *Sustainable Materials and Technologies*, 2025. **43**.
58. Palanivel, P.D., *et al.*, Harnessing dark fermentation: Mechanisms, metabolic pathways, and nanoparticle innovations for biohydrogen enhancement. *International Journal of Hydrogen Energy*, 2025. **109**: p. 383–396.
59. Carvalho Miranda, J.C.D., *et al.*, The landscape of green and biohydrogen technology: A data-driven exploration using non-supervised methods. *International Journal of Hydrogen Energy*, 2025. **105**: p. 1467–1490.
60. Mozas Santhose Kumar, J., R. Prakash, and P. Panneerselvam, Hydrothermal liquefaction: A sustainable technique for present biofuel generation: Opportunities, challenges and future prospects. *Fuel*, 2025. **385**.
61. Rojas, M., *et al.*, Advances and challenges on hydrothermal processes for biomass conversion: Feedstock flexibility, products, and modeling approaches. *Biomass and Bioenergy*, 2025. **194**.
62. Usman, M., *et al.*, From biomass to biocrude: Innovations in hydrothermal liquefaction and upgrading. *Energy Conversion and Management*, 2024. **302**.
63. Dsilva Winfred Rufuss, D., K.S. Sonu Ashritha, and L. Suganthi, Green energy revolution: A unique approach for energy forecasting and optimization towards sustainable energy planning and social development. *Environment, Development and Sustainability*, 2024.
64. Chaudhary, V.P., *et al.*, Agri-biomass-based bio-energy supply model: An inclusive sustainable and circular economy approach for a self-resilient rural India. *Biofuels, Bioproducts and Biorefining*, 2022. 16(5): p. 1284–1296.
65. Rahimpour, M.R., *Encyclopedia of Renewable Energy, Sustainability and the Environment*. Vol. 1–4. 2024: Elsevier. 1–3899.

66. Rahimpour Golroudbary, S., M. Lundström, and B.P. Wilson, Synergy of green energy technologies through critical materials circularity. *Renewable and Sustainable Energy Reviews,* 2024. **191.**

67. Vicidomini, M., Renewable energy systems 2020. *Applied Sciences (Switzerland),* 2021. **11**(8).

68. Rządkowska, A.E., Quantitatively estimating the impact of the European Green Deal on the clean energy transformation in the European Union with a focus on the breakthrough of the share of renewable energy in the electricity generation sector. *Polityka Energetyczna,* 2022. **25**(2): p. 45–66.

69. Chen, W. and M. Han, Quantifying the cost savings of global solar PV and onshore wind markets. *Energy Policy,* 2025. **199.**

70. Victoria, M., *et al.*, Solar photovoltaics is ready to power a sustainable future. *Joule,* 2021. **5**(5): p. 1041–1056.

71. IEA, *World Energy Outlook 2024,* P. IEA, Editor. 2024.

72. Mohanty, S., *et al.*, Optical power monitoring systems for offshore wind farms: A literature review. *Sustainable Energy Technologies and Assessments,* 2024. **72.**

73. Mohanty, V., *et al.* Sustaining scalable sustainability: Human-centered green technology for community-wide carbon reduction, in *Conference on Human Factors in Computing Systems - Proceedings.* 2024. Association for Computing Machinery.

74. Haces-Fernandez, F., M. Cruz-Mendoza, and H. Li, Onshore wind farm development: Technologies and layouts. *Energies,* 2022. **15**(7).

75. Lorca, Á., M. Favereau, and D. Olivares, Challenges in the management of hydroelectric generation in power system operations. *Current Sustainable/Renewable Energy Reports,* 2020. **7**(3): p. 94–99.

76. Rahman, A., O. Farrok, and M.M. Haque, Environmental impact of renewable energy source based electrical power plants: Solar, wind, hydroelectric, biomass, geothermal, tidal, ocean, and osmotic. *Renewable and Sustainable Energy Reviews,* 2022. **161.**

77. Gemechu, E. and A. Kumar, A review of how life cycle assessment has been used to assess the environmental impacts of hydropower energy. *Renewable and Sustainable Energy Reviews,* 2022. **167.**

78. Osman, A.I., *et al.*, Cost, environmental impact, and resilience of renewable energy under a changing climate: A review. *Environmental Chemistry Letters,* 2023. **21**(2): p. 741–764.

79. Finecomess, S.A. and G. Gebresenbet, Future green energy: A global analysis. *Energies,* 2024. **17**(12).

80. Tester, J.W., *et al.* Sustainable energy: Choosing among options, in *AIChE Annual Meeting, Conference Proceedings.* 2005. American Institute of Chemical Engineers.

81. Chaudhary, A., *et al.*, Recent technological advancements in biomass conversion to biofuels and bioenergy for circular economy roadmap. *Renewable Energy,* 2025. **244.**

82. Zhou, Z., *et al.*, A review of energy storage technologies for marine current energy systems. *Renewable and Sustainable Energy Reviews,* 2013. **18**: p. 390–400.

83. Zhang, W., *et al.*, Harvesting energy from marine: Seawater electrolysis for hydrogen production. *Fuel,* 2024. **377.**

84. Trunina, I., K. Pryakhina, and S. Yakymets. Research on the development of renewable energy sources in the world due to the war in Ukraine, in *Proceedings of the 2022 IEEE 4th International Conference on Modern Electrical and Energy System, MEES 2022*. 2022. Institute of Electrical and Electronics Engineers Inc.
85. Khan, A., T. Khan, and M. Ahmad, The role of technological innovation in sustainable growth: Exploring the economic impact of green innovation and renewable energy. *Environmental Challenges*, 2025. **18**.
86. Pagliaro, M. and F. Meneguzzo, Distributed generation from renewable energy sources: Ending energy poverty across the world. *Energy Technology*, 2020. 8(7).

Chapter 8

GSE harnessing, harvesting, and storage

Everything is Energy and that is all there is to it. Match the frequency of the reality you want and you can not help but get that reality. It can be no other way. This is not philosophy. This is physics.

Albert Einstein [1]

8.1 INTRODUCTION

Knowing there is an abundance of forces and energies in the Earth's environment, the challenge is now to capture it effectively and efficiently (see Figure 8.1). It must be in a form that is both usable and able to be stored. In this chapter we will explore energy harnessing and harvesting, which will require storage. Both processes are being used to capture energy from the environment and convert it into electrical power, which can then be

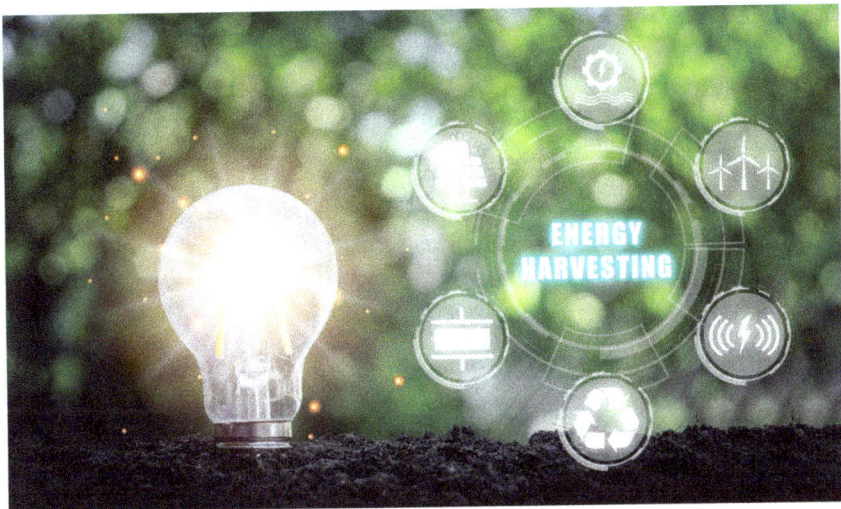

Figure 8.1 Harvesting and harnessing energy.

Source: Shutterstock Photo ID: 2322074783.

DOI: 10.1201/9781003407447-11

used to supply needed energy to the world. Harnessing is often used in reference to powerful energy sources that supply large amounts of energy that serve daily use, such as solar or hydroenergy. Harvesting refers to the gathering of ambient energy that tends to be continuously generated over time. For example, a factory could use devices that generate electricity from mechanical vibration (usually piezoelectric devices); this would be energy harvesting. With respect to GSE sources, both "harness" and "harvest" are used to describe the capture of usable energy from the environment. As will be seen, they result in the production of electricity, which is becoming the predominant form of energy in our world.

GSE harnessing and harvesting are key emerging technologies of the 21st century. The choice of using harnessing or harvesting to acquire GSE will depend on the immediate application or long-range goal, size, and environment. Different GSE sources have different strengths and weaknesses, and some may be more suitable than others depending on the circumstances [2]. Harnessing/harvesting corresponds to two scales of application: large and small. In line with green and sustainability principles, the collection of energy from the environment also recovers energy that would otherwise be lost in terms of heat (circularity). GSE harnessing/harvesting will collect local naturally available energy from the environment for local use or larger continuous collections that will be distributed or stored.

GSE sources different from traditional fossil energy are being developed, including salinity gradient energy, hydroenergy, and thermal conversion, to name a few [3]. As can be seen, there are numerous sources of sustainable energy production, which, when combined, will help meet the increasing world demand for energy. The future prospects and potential direction of these emerging technologies are important for GSE [4].

8.2 HARVESTING AND HARNESSING GSE

The global energy transition to green sustainable energy is a defining challenge of our era. Its success depends on more than mere environmental sentimentality, but it will rely heavily on science and technology (see Figure 8.2). The transition to GSE must address three diverse areas:

- Global and inclusive addressing the diverse needs of geographic regions.
- Development of innovative mechanisms to draw from GSE sources and to scale technologies effectively.
- Recognizing the specific challenges and opportunities for small-, medium-, and large-sized energy recovery.

GSE sources (see Chapter 7, "Sources of green sustainable energy") offer the means to facilitate the emerging energy transition (see "Epilogue: The

Figure 8.2 Harnessing GSE energy.

Source: Shutterstock Photo ID: 1674291655.

role of GSE in energy transition"). These sources must be coupled with green and sustainable technologies to guide the harnessing/harvesting efforts to meet global sustainability goals. GSE sources, including hydro, geothermal, biomass, solar, and wind energy, are developed and marketed as low- or non-carbon alternatives to conventional energy sources, but technologies used to capture and store them must be able to meet the demand. The development of green sustainable energy technologies will allow GSE to be recovered in a sustainable manner while cutting down on waste and greenhouse gas emissions, thus enhancing the overall carbon footprint of energy production [5].

8.2.1 Energy distribution in the 21st century

In the global transition toward GSE, there are significant disparities in adoption rates and technological advancements across nations. This probably will diminish as the effectiveness and availability of GSE increases. This points to the potential for an extensive shift in energy paradigms. A recent study indicates that renewable energy sources could entail up to two-thirds of the global primary energy supply by 2050, and renewable energies should develop into GSE. Led by the European Union countries, countries in Asia and the Americas are integrating new energy technologies at varying rates. Meanwhile, Middle Eastern countries are progressively diversifying their energy portfolios, and Africa is potentially open to these trends [6]. Energy use continues to grow globally, and decarbonizing the energy system must be undertaken carefully to ensure that the global economy is not damaged. The pace of change is not rapid, but it is steady.

A significant component of this transition is a significant shift from centralized energy systems to decentralized ones. "Centralized generation" refers to the large-scale generation of electricity at centralized facilities. These facilities are usually located away from end-users and connected to a network of high-voltage transmission lines. The electricity generated by centralized generation is distributed through the electric power grid to multiple end-users. Centralized generation facilities include fossil-fuel-fired power plants, nuclear power plants, hydroelectric dams, wind farms, and more [7]. The use of more GSE will require decentralizing the energy distribution system, and this becomes a new paradigm. These decentralized systems aim to satisfy local energy needs using GSE resources within the community. This approach leads to decreased complexity and costs, improved efficiency, and enhanced local resilience and supports energy independence, thereby advancing the transition toward zero carbon emissions [8].

Electricity is the form of all energy that is harnessed/harvested; it is most useful in this form. Using electricity, a decentralized system relies on the following characteristics, which will define the GSE distribution system [9]:

- Distributed generation: decentralized power generation allows coordination between heat and power generation which increases the system's efficiency.
- Energy storage: decentralized power adds more generation sources which can lead to new storage options, which is particularly helpful for supplementing energy plants, which often produce during non-peak hours.
- Demand response: New technologies will include smart grid and smart metering which allow for real-time monitoring of energy needs and communication between producers and consumers of electricity to optimize grid usage.
- Infrastructure design: Distributed generation facilities may be connected to the grid or simply serve a particular site without necessarily feeding potential excess generation into the grid.

8.2.2 Energy form and units

The need for energy continues to grow as the population expands. The sources for that energy are changing due to the global demand to address climate change. Globally we get the largest amount of our energy from oil, followed by coal, gas, and hydroelectric power. This volume is determined by the energy needs of the world. GSE and other renewable sources must meet these growing energy needs [10].

There are five energy-use sectors: electric power, transportation, industrial, residential, and commercial. Total energy consumption by the end-use sectors includes their primary energy use, purchased electricity, and also electrical system energy losses (energy conversion and other losses associated with the generation, transmission, and distribution of purchased electricity), and other energy losses. The energy used by each sector can be met by the variety of available GSE through harnessing, harvesting, and subsequent storage.

Existing energy sources are measured in different physical units: liquid fuels in barrels or gallons, natural gas in cubic feet, coal in short tons, and electricity in kilowatts and kilowatt-hours. In the United States, the British thermal unit (Btu), a measure of heat energy, is commonly used for comparing different types of energy to each other. GSE will mostly employ liquid fuels in barrels or gallons and electricity in kilowatts and kilowatt-hours as units of measure in the five energy use sectors. It will also retain the common thermal units.

The GSE that is harnessed is mostly converted to electricity. In this form it is distributed by electricity providers. The GSE sources include solar, water, and wind. Electricity is becoming the common form for most of the general energy distribution. This is the current path of energy transition.

GSE that are harvested are smaller, and they are also in the form of electricity. Health, infrastructure, and environmental monitoring as well as networking and defense technologies are only some of the potential areas of application of these systems. It is highly desirable that they operate without an external electricity source and instead draw the energy they require from the environment in which they are used [11].

8.2.3 Energy harnessing/harvesting

Today's changing energy landscape does not favor any single source to provide the entire solution to the world's energy challenges. Collaborative innovation – the partnering of different technologies across the environmental and electromagnetic networks will facilitate the development and introduction of new and necessary energy solutions. Working with nature through harnessing and harvesting allows for meeting energy needs while preserving and saving the Earth's environment.

8.2.3.1 Harnessing/harvesting within energy production

Energy production is a way to describe the generation of energy from green and renewable sources such as bioenergy, solar energy, geothermal energy, wind energy, and hybrid energy systems, which do not rely on fossil fuels and contribute to the stability and preservation of the ecosystem [12]. This energy production can be broken down into harnessing and harvesting energy. As will be seen in the following section, these two processes more appropriately describe different scales of energy produced.

Energy harnessing involves capturing available energy and converting it to electrical power. Usually, this harnessing is on a large scale, and it contributes to our global energy needs by adding to the energy grid. Energy harvesting, on the other hand, is a process by which energy is derived from ambient background energy sources. Energy harvesting usually provides a very small amount of power for low-energy electronics. The line of separation between these two descriptions is not so strict that harnessing and harvesting are sometimes used interchangeably.

More importantly, these ways of capturing environmental energies are forming the backbone of the energy transition (see Chapter "Epilogue: The role of GSE in energy transition").

8.2.3.2 Size of energy recovery

The inevitable transition of the energy sector from using conventional sources (fossil fuels) to sustainable sources raises serious concerns and questions regarding both feasibility and reliability. If they are to replace existing energies, there will be systemic changes. These new energy sources will need to supply energy in varying quantities, depending upon the application. New energy sources will need to meet the established standards, which are defined by the form/application, distribution potential, and the amount required. New emerging energy sources will include large amounts like wind/solar/geothermal. There will also be the by-products of energy systems after useful conversion such as discharged heat from turbine and exhaust flue gases (usually never considered as a source but rather referred to as energy inefficiencies) [13]. For GSE a valuable way to characterize the sources and energy recovery is by size (shown in Tables 8.1 and 8.2). This will allow the sources can be summed to ascertain if they will meet demands and the sustainability requirement goals for energy appropriate to the 21st century.

Table 8.1 Large (usually energy harnessing): Existing energy sources, estimated potential, and energy density [13]

Energy source	Useful potential at present
Solar	12,000 EJ (EJ) per year
Wind (on and offshore)	98,000 EJ per year
Tidal	800 TW-hour (TWh) per year
Wave	2000–4000 TWh per year
Biomass (vary with the material)	160–200 EJ per year
Geothermal	53,000 EJ per year

Table 8.2 Small (usually energy harvesting): Representative ambient harvested energies and related power densities [14]

Technologies	Power density
Radiofrequency	[1.2.10−5−12]; mW/cm^2
Thermoelectric	[15−60]; µW/cm^3
Piezoelectric	[0.11−7.31]; mWg^2/cm^3
Photovoltaic	[0.006−15]; mW/cm^2

8.3 ENERGY HARNESSING

The energy contained in the Earth's environment is massive. Energy harnessing captures available energy and converts it to electrical power. Energy sources can be harnessed in many ways, including capturing solar, wind, and water energy. These are occurring simultaneously with the more prevalent ones using fossil fuels (oil, coal, and gas). The harnessed energy is converted into mechanical energy, then to electrical energy, to use it for other purposes. There are many ways in which we harness nature's energies, but with GSE this harnessing is accomplished with little or no impact on the environment (see Figure 8.3).

The revolutionary possibilities of machine learning (ML) and intelligent algorithms for enabling green sustainable energy, with an emphasis on

Figure 8.3 Need for energy harnessing in a green world.

Source: Shutterstock Photo ID: 2289699573.

the energy domains of solar, wind, biofuel, and biomass are adding new weight and emphasis to using these alternative energies. Critical problems with harnessing (such as data variability, system inefficiencies, and predictive maintenance) can be addressed by the integration of ML in renewable energy systems. As one example, machine learning improves solar irradiance prediction accuracy and maximizes photovoltaic system performance in the solar energy sector. There is a vital role of ML in promoting harnessing in sustainable and efficient GSE systems [15].

8.3.1 Sources and technologies for harnessing

Sources and technologies associated with harnessing GSE are well known but also are expanding. To meet the growing demand for clean, sustainable, and viable energy in the 21st century traditional sources like solar power, hydropower, and wind energies can offer the most immediate and viable alternatives to meet future energy demands [16]. In this section are listed sources and technologies that are harnessed to produce GSE.

8.3.1.1 Solar power

The sun provides the most abundant source of energy. Solar power ranks first among the GSE energy sources. Using solar panels to convert light into electricity, this abundant energy source varies based on geographic location but has the potential to be harnessed almost everywhere. From vast solar farms in sunny rural areas to rooftop installations in urban settings, solar energy is a cornerstone of renewable energy systems [17].

8.3.1.2 Wind power

Wind is abundant in many places and is green and sustainable. Wind is harnessed using giant wind turbines that rotate in high winds, converting the energy into the kinetic energy of wind into mechanical and then electrical energy. Using wind turbines strategically placed in high-wind areas, often in coastal or open plain regions, wind farms can generate significant amounts of electricity [18].

8.3.1.3 Water and wave energy

Flowing rivers provide another abundant source of energy. Capturing it requires hydroelectric plants and dams that are expensive but can pay for themselves over time. The dams and plants, much like wind turbines, generate electrical energy from mechanical motion. Water flow, from rivers to man-made reservoirs, can also produce energy. This type of power is particularly beneficial in areas with high levels of rainfall and flowing water, and its generation can be adjusted based on electricity demand, offering a reliable and flexible renewable energy source.

While not as widely recognized as hydroelectric, other renewable sources like tidal and wave energy, which harness the power of ocean currents, are gaining attention. These emerging technologies have the potential to play a significant role in our future energy mix, especially in coastal regions [19].

8.3.1.4 Geothermal

A valuable green sustainable energy source is heat from the Earth's crust. Geothermal energy sources use the steam from heated groundwater to generate electricity. This heat, derived from the Earth's hot magma and radioactive particles within the crust, provides a powerful and consistent source of energy, especially beneficial in regions with volcanic activity or hot springs [20].

8.3.1.5 Biomass

Biomass energy is produced from organic material, such as plant matter and animal waste, which can be burned or biochemically processed to produce energy. This source is not totally zero CO_2, so it should not become the only solution. It can contribute to energy circularity. This type of energy is especially relevant in rural areas and places with abundant organic matter, such as agricultural waste. While biomass is renewable, it must be managed sustainably to prevent negative impacts on the environment [21].

8.3.2 Examples of harnessing

Table 8.3 lists some examples of energy harnessing.

Table 8.3 Examples of GSE harnessing

Energy source	Technologies	References
Solar power	• Solar panels on homes and office buildings can capture energy.	[22, 23]
Wind power	• Multi-energy wind power plants can produce electricity, fresh water, hydrogen, and oxygen.	[24, 25]
Geothermal	• Capturing geothermal heat with water and then transforming that heat into electricity.	[5, 26–28]
Biomass	• Biomass feedstock (derived from organic materials like agricultural waste and forestry wood) recycles CO_2 (through processes like gasification, pyrolysis, and hydrothermal liquefaction).	[16, 29, 30]
Water and wave energy	• Hydropower comprises the largest portion of the world's current GSE generation.	[31–33]

Figure 8.4 Solar thermal energy collectors.
Source: Shutterstock Photo ID: 1802214730.

8.4 ENERGY HARVESTING

Energy harvesting is the collection of ambient energy from the environment; energy that would otherwise be lost to heat. Energy harvesting is basically the collection of local naturally available energy for local use. The energy scale is small, but nonetheless not insignificant (see Figure 8.4). The main category of applications at these power levels is wireless devices. The applicability of energy harvesting to devices depends on the type and amount of the available ambient energy as well as on size limitations. Thermal harvesters, for example, use the thermoelectric effect, and light harvesters the photoelectric effect, while electromagnetic harvesters use induction. Chemical harvesters can employ a variety of chemical reactions on the surface of electrodes [34].

8.4.1 Sources and technologies for harvesting

With minimal environmental effects, energy savings, and sustainability, the technology of energy harvesting and conversion is important. Energy harvesting utilizes responsive materials and corresponding devices recently developed for the generation of different kinds of energies (such as electricity and chemical energy, and particularly hydrogen energy) by energy harvesting. Electrical energy can be generated based on four effects (thermoelectric, electrokinetic, triboelectric, and hydrovoltaic) and two processes (electrochemical and photoelectric chemical), while hydrogen can be produced based on photoelectrochemistry, photocatalysis, and photovoltaic electrolysis [35].

8.4.1.1 Relative motion

Mechanical energy harvesters based on smart and multifunctional materials, such as piezoelectric materials, pyroelectrics, and ferroelectric materials, are used to capture energy from motion. They have been developed as potential energy transducers to replace or supplement batteries. These are able to fill an energy need with a GSE solution [36].

8.4.1.2 Temperature gradient

Thermoelectric materials allow direct energy conversion without moving parts and do not produce greenhouse gas emissions, employing lightweight and quiet devices [37]. More specifically, thermal energy harvesting converts the thermal energy collected from a heat source into electricity.

8.4.1.3 Light

A green energy approach is to harness efficiently solar energy into green chemical and electrical energy. The sunlight energy cannot be utilized directly as a result of its intermittent and diffuse nature, so there is the development of efficient and economically viable technologies for its efficient conversion to an applicable source of energy [38].

8.4.1.4 Chemical energy

The formation of chemical compounds allows the storage of energy. Using renewable sources, such as solar and wind, can produce electrical energy. A viable route for storing this electrical energy at a massive scale is its conversion into chemical energy carriers by combining or integrating electrochemistry with biology and chemical engineering [39].

8.4.1.5 Microbial fuel cell (MFC)

The development of a microbial fuel cell (MFC) makes it possible to generate clean electricity as well as remove pollutants from wastewater. Extensive studies on MFC have focused on structural design and performance optimization, and tremendous advances have been made in these fields. MFC-based wastewater treatment and energy harvesting biocatalysts prove to be very useful. This is simultaneously a wastewater treatment and sustainable energy harvesting [32].

8.4.1.6 Salinity gradient energy harvesting

Harvesting salinity gradient energy, also known as "osmotic energy" or "blue energy," results in energy being generated from the free energy mixing of seawater and fresh river water. The blue energy is in the form of chemical energy.

8.4.2 Examples of harvesting

Table 8.4 lists technologies that can be found in each of the areas listed above.

Table 8.4 Examples of GSE harvesting

Energy source	Technologies	References
Relative motion	• Piezoelectric mechanical energy harvesting – converts human motion into electricity.	[40, 18]
	• Electromagnetic forces – convert mechanical energy into electric energy both for large-scale and small-scale applications.	[41]
	• Electrostatic forces – arise from the charge of two interacting surfaces for use in embedded systems and wireless sensor nodes.	[42, 43]
Temperature gradient	• Thermoelectric forces – harvesting electrical energy by transforming heat energy.	[44, 45]
Light	• Photoelectric – scavenging energy using photoelectric, thermoelectric, electrostatic, and electromagnetic conversion techniques is increasingly possible.	[46]
	• Electromagnetic radiation – capturing energy from solar energy in the form of electromagnetic waves such as radio waves, visible light, and gamma rays.	[47]
	• Harvesting and detection – electromagnetic (EM) wave harvesting and detection used for self-powered human health monitoring and robot intelligence.	[48]
	• Radiofrequency energy harvesting – offers the supply of wireless power for battery-free devices.	[49, 50]
Chemical energy	• Chemical reactions – utilizing catalytic materials for the photo- and photo-electro-catalytic water splitting, photovoltaic cells, and fuel cells.	[51]
	• Harvesting salinity gradient energy, also known as "osmotic energy" or "blue energy," is generated from the free energy mixing of seawater and fresh river water.	[52, 53]
Microbial fuel cell (MFC)	• Microbial fuel cell – electricigens in wastewater can act as catalysts for destroying organic pollutants to produce bioelectricity.	[54]

8.5 SUSTAINABLE ENERGY STORAGE

Transformation of energy supply systems into green sustainability expands the variety of renewable energy sources. Renewable sources cannot continuously supply energy. Therefore, energy storage systems are vital to the success of the whole system of generation and distribution. Energy storage systems have many issues in terms of sustainability. Key performance indicators define the sustainability of energy storage systems. The least best impacts occur in the performance of mechanical energy storage and sensible/latent heat storage. The production of green hydrogen, green ammonia, and biogas shows some negative impact. The worst sustainability is found in energy storage technologies for electrochemical energy storage technologies [55].

On smaller scales, storage of energy is essential to meet the daily demand for powering portable devices (Figure 8.5). This necessitates the development of storage systems such as supercapacitors (SCs), batteries, and solar cells. SCs have gained significant attention for their ability to provide a massive amount of power. Nevertheless, traditional mechanisms fall short

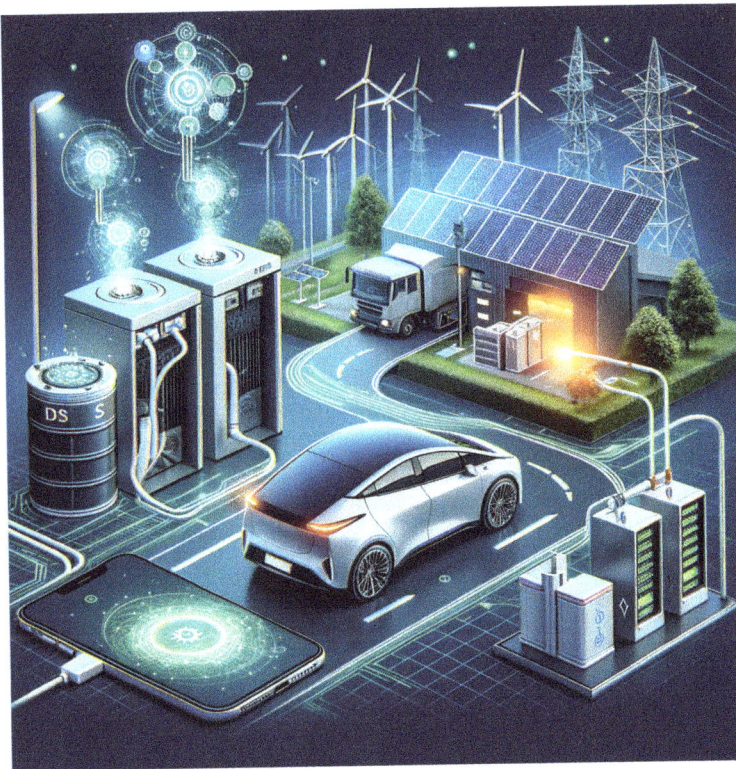

Figure 8.5 GSE storage.

Source: Shutterstock Photo ID: 2408167879.

of expectations [56]. The main advantages of supercapacitors are their light weight, volume, greater life cycle, turbocharging/discharging, high energy density and power density, low cost, easy maintenance, and no pollution.

Current energy storage systems (ESS) have the disadvantages of self-discharging, low energy density, poor life cycles, and high cost. Ambient energy resources are the best option as an energy source, but the main challenge in harvesting energy from ambient sources is the instability of the source of energy. Due to the explosion of lithium batteries in many applications, and the increased effectiveness associated with them, the design of new efficient devices, which are more reliable and efficient than conventional batteries, is important [57].

8.5.1 Need for energy storage

The rising demand for green energy to reduce carbon emissions is accelerating the integration of renewable energy sources (RESs). However, this shift presents significant challenges due to the inherent variability and intermittency of RESs, which does impact power system stability and reliability. As a result, there is a simultaneous need for enhanced storage to maintain stable and reliable operations. Recent challenges, such as the increased risk of grid congestion, frequency fluctuations and deviations, and the need for real-time supply and demand balancing, necessitate innovative ESS applications. This is driving a transition pathway to promote the large-scale deployment of diverse ESS technologies to support grid modernization, enhance resilience, and foster sustainable power supply development [58].

8.5.2 Scale

The demand for renewable energy sources worldwide has pushed the need for more storage. The economical use of these technologies has been researched but has not been widely adopted due partly to cost and the inability for continuous and reliable service during off-peak periods. To make these technologies more competitive, research into energy storage systems has intensified over the last few decades. The goal is to devise an energy storage system that allows for storage of electricity during lean hours at a relatively cheaper value and delivery later. Energy storage and delivery technologies such as supercapacitors can store and deliver energy at a very fast rate, offering high current in a short duration [59].

On smaller scales, organic electrode materials (OEMs) can deliver remarkable battery performance for metal-ion batteries (MIBs) due to their unique molecular versatility, high flexibility, versatile structures, sustainable organic resources, and low environmental costs. This highlights how molecular engineering can offer practical paths for developing advanced OEMs that can be applied in next-generation rechargeable MIBs [60].

8.5.3 Storage capacity

Supercapacitors are emerging as a pivotal technology as they provide quick charge/discharge rates and act as a bridge between batteries and conventional capacitors. Supercapacitors have the potential to address the pressing challenges associated with energy storage and pave the way for more sustainable and responsive energy systems. Supercapacitors fill the gap between batteries and capacitors in terms of energy density and power density and open a wide field for applications [61].

Electrochemical energy storage devices, particularly supercapacitors, have found applications in electric vehicles, power supports, portable electronics, and many other applications requiring electric energy devices for their operation. Thus, the growth of these SCs in the commercial market has met requirements, but further developments are necessary for their continued effective industrialization. In the meantime, SCs also face technical complications and hurdles for their deployment in industrial settings because of their low energy density and high cost [62].

8.5.4 New developments

Harnessing biomass to fabricate electronic and storage devices is an example of novel research on storage. It not only represents a promising strategy for making materials but is also beneficial for the sustainable development of technologies. Numerous recent studies have demonstrated that green materials can be promising candidates for synthesizing high-performance materials, such as biopolymers and hierarchical porous carbons. As the state-of-the-art green materials advance, recent progress in biodegradable polymeric materials and biomass-based carbon materials, together with their electronics and energy storage applications, provides new avenues for energy. This blends nicely with the merits of integrating green design with GSE [63].

8.5.4.1 Electrochemical

With the energy crisis and environmental pollution, the development of sustainable new energy has become an urgent priority. Society is aware that green energy technologies are critical to economic development. Electrochemical energy storage technology is a green energy technology based on their ability to provide high energy density (battery) or high-power density (supercapacitor). There is a growing demand for high-energy and high-power-density electrochemical energy storage devices for current and future applications. Research has found that new semiconductor materials have the potential to improve the cycle life and energy and power density of supercapacitors. To date, a variety of novel semiconductor electrode materials have been fabricated and studied for supercapacitors [64].

8.5.4.2 Hybrid systems

Hybrid systems that integrate solar cells and supercapacitors have gained significant attention among researchers and scientists worldwide. This has helped to fuel demands for GSE, miniaturization, and mini-electronic wearable devices. These hybrid devices will lead to sustainable energy becoming viable and fossil-fuel-based sources of energy gradually being replaced. A solar photovoltaic (SPV) system is an electronic device that mainly functions to convert photon energy to electrical energy using a solar power source [16]. These hybrids will be increasingly important during the period of energy transition.

8.5.4.3 Batteries

Technological advancements in electric vehicles (EVs), especially in the realm of storage technology, battery management systems, power electronics technology, charging strategies, methods, algorithms, and optimizations, are aiding the progress of energy storage. EVs influence various goals of sustainable development, such as affordable and clean energy, sustainable cities and communities, industry, economic growth, and climate actions. EV technology is beneficial to GSE as it influences the development of efficient battery storage, charging approaches, converters, controllers, and optimizations toward targeting SDGs [65].

Modern lifestyle is pushing the development of energy storage technology for the maximum utilization of sustainable green energy. Lithium-ion batteries (LIBs) are critical owing to their use in a variety of electronic devices as rechargeable energy storage devices. New anode material for (LIBs) is under study, but several obstacles remain [66].

The next-generation batteries will incorporate climate-neutral energy principles. A new area of research, the biofuel cell-based biobattery, is a net-zero better alternative to conventional biofuel cells, although this class of biobatteries is still in the development stage [67].

8.5.4.4 Rechargeable battery systems

Decarbonization of the grid requires the increased development of rechargeable energy storage. The electrochemical battery systems are a vital transformative technology that will support GSE power through efficient storage. However, improvements are needed in both these and in the commercial Li-ion batteries (LIBs) technologies to meet the need for rechargeable battery systems [68].

8.5.5 Modern energy storage

Energy storage and environmental protection are the major challenges of the 21st century, the world has to face. Extensive efforts have been made to develop Electrochemical Energy Storage (EES) devices like supercapacitors based on green and sustainable principles. Nano Cellulose (NC) is being developed as a sustainable nanomaterial with its unique structure and properties, such as high specific modulus, excellent stability in most solvents, low toxicity, and natural abundance. Low cost and simple synthesis techniques further enable the NC as a promising alternative material for the fabrication of renewable energy storage devices. NC-based electrodes, separators, and electrolytes as supercapacitor components are areas of great promise [69].

Energy storage is essential to the success of many electrical and electronic applications powered through GSE. Employing a hybrid functionality of harvesting and storage provides a pathway to new technologies. There are many challenges in the merging of these two dimensions of GSE. Successes are seen now in low-voltage and low-power electronics and wearable electronics [70].

8.5.6 Future

The increased usage of renewable energy sources (RESs) and the intermittent nature of the power they provide highlight issues of stability, reliability, and power quality. In such instances, energy storage systems offer a promising pathway to more GSE adoption. These issues are not adequately resolved, as a single ESS will not fulfill all the requirements for all operations. Due to the limited capability of a single ESS and the uncertainty concerning cost, lifespan, and power and energy density, there will be ample opportunities for research. To overcome the trade-off issue resulting from using a single ESS system, a hybrid energy storage system (HESS) consisting of two or more ESSs appears as an effective solution. Many studies have been considered lately to develop and propose different HESSs for different applications showing the great advantages of using multiple ESSs in one combined system. The emerging role of HESSs in terms of their benefits and applications is important to pursue. Recent control and optimization methods of HESSs associated with RESs and their advantages and disadvantages have been reviewed. Hopefully, increased efforts will lead to the development of an advanced HESS for future renewable energy optimal operation [71].

8.6 CONCLUSION

The effective use of harnessing and harvesting technologies, coupled with increased storage capacity, is expanding GSE use. With the higher demand for energy storage device performance, supercapacitors have attracted increasing interest because of their high-power density, stable cycling capability, and wide range of operating temperatures. To address the worsening climate change, biomass, as a green renewable resource, is providing a transition to lower/zero-carbon sources. Similarly, biochar, with its naturally porous structure and abundant surface functional groups, is considered a promising candidate for the development of energy storage [72]. It is significant that controllable production for improving the electrochemical performance of supercapacitors and the recent research achievements of GSE sources and capacitors will provide strong support for the eminent energy transition.

REFERENCES

1. Einstein, A., *Albert Einstein: The Human Side*. 1981: Princeton University Press.
2. Kilner, J.A., *et al.*, Functional materials for sustainable energy applications, in *Functional Materials for Sustainable Energy Applications*. 2012: Elsevier Ltd. 1–681.
3. Fang, Z., *et al.*, Nanochannels and nanoporous membranes in reverse electrodialysis for harvesting osmotic energy. *Applied Physics A: Materials Science and Processing*, 2022. **128**(12).
4. Bagherzadeh, R., *et al.*, Wearable and flexible electrodes in nanogenerators for energy harvesting, tactile sensors, and electronic textiles: Novel materials, recent advances, and future perspectives. *Materials Today Sustainability*, 2022. **20**.
5. Sharma, V.K., *et al.*, A comprehensive review of green energy technologies: Towards sustainable clean energy transition and global net-zero carbon emissions. *Processes*, 2025. **13**(1).
6. Hassan, Q., *et al.*, The renewable energy role in the global energy transformations. *Renewable Energy Focus*, 2024. **48**.
7. Ren, D. and X. Guo, The economic use of centralized photovoltaic power generation: Grid connection, hydrogen production or energy storage? *Journal of Energy Storage*, 2025. **106**.
8. Ahmed, S., A. Ali, and A. D'Angola, A review of renewable energy communities: Concepts, scope, progress, challenges, and recommendations. *Sustainability (Switzerland)*, 2024. **16**(5).
9. Weinand, J.M., F. Scheller, and R. McKenna, Reviewing energy system modelling of decentralized energy autonomy. *Energy*, 2020. **203**.
10. Ritchie, H. and P. Rosado *Energy Mix*. 2020. (Published online at OurWorldinData.org.)
11. Wang, Z.L. and W. Wu, Nanotechnology-enabled energy harvesting for self-powered micro-/nanosystems. *Angewandte Chemie - International Edition*, 2012. **51**(47): p. 11700–11721.
12. Ma, Z., M. Arıcı, and A. Shahsavar, Building Energy Flexibility and Demand Management, in *Building Energy Flexibility and Demand Management*. 2023: Elsevier. 1–270.
13. Alajingi, R. and M. R, Novel classification of energy sources, with implications for carbon emissions. *Energy Strategy Reviews*, 2023. **49**.

14. Mohammadi, M. and I. Sohn, AI based energy harvesting security methods: A survey. *ICT Express*, 2023. **9**(6): p. 1198–1208.
15. Le, T.T., *et al.*, Unlocking renewable energy potential: Harnessing machine learning and intelligent algorithms. *International Journal of Renewable Energy Development*, 2024. **13**(4): p. 783–813.
16. Yadav, P.K., A. Kumar, and S. Jaiswal, A critical review of technologies for harnessing the power from flowing water using a hydrokinetic turbine to fulfill the energy need. *Energy Reports*, 2023. **9**: p. 2102–2117.
17. Abidin, N.I., *et al.* The determinants factors for solar photovoltaic implementation in existing building, in *Key Engineering Materials*. 2022. Trans Tech Publications Ltd.
18. Spiru, P., Assessment of renewable energy generated by a hybrid system based on wind, hydro, solar, and biomass sources for decarbonizing the energy sector and achieving a sustainable energy transition. *Energy Reports*, 2023. **9**: p. 167–174.
19. Androniceanu, A. and O.M. Sabie, Overview of green energy as a real strategic option for sustainable development. *Energies*, 2022. **15**(22).
20. Herrera-Franco, G., *et al.*, Bibliometric analysis and review of low and medium enthalpy geothermal energy: Environmental, economic, and strategic insights. *International Journal of Energy Production and Management*, 2023. **8**(3): p. 187–199.
21. Rojas, M., *et al.*, Advances and challenges on hydrothermal processes for biomass conversion: Feedstock flexibility, products, and modeling approaches. *Biomass and Bioenergy*, 2025. **194**.
22. El-kenawy, E.S.M., *et al.*, Global scale solar energy harnessing: An advanced intra-hourly diffuse solar irradiance predicting framework for solar energy projects. *Neural Computing and Applications*, 2024. **36**(18): p. 10585–10598.
23. Cai, S., *et al.*, Harnessing solar energy by a self-driven photoelectrocatalytic system for versatile water purification: Radionuclides, organic pollutants and pathogen removal. *Separation and Purification Technology*, 2025. **362**.
24. Ahern, C., R. Oliver, and B. Norton, Harnessing curtailed wind-generated electricity via electrical water heating aggregation to alleviate energy poverty: A use case in Ireland. *Sustainability (Switzerland)*, 2024. **16**(11).
25. Assareh, E., *et al.*, Integrated wind farm solutions: Harnessing clean energy for electricity, hydrogen, and freshwater production. *International Journal of Hydrogen Energy*, 20241213-1227)
26. Kata, D., J. Gatune, and I. Kanana. Harnessing geothermal energy for decarbonization and sustainable agricultural development using Geotto, in *Advances in Science and Technology*. 2025. Trans Tech Publications Ltd.
27. Sarsenbayev, D., *et al.*, Harnessing geothermal energy potential from high-level nuclear waste repositories. *Energies*, 2024. **17**(9).
28. Liu, X., G. Falcone, and C. Alimonti, A systematic study of harnessing low-temperature geothermal energy from oil and gas reservoirs. *Energy*, 2018. **142**: p. 346–355.
29. Kumar, A., *et al.*, From waste to fuel: Harnessing high specificity lipases from Candida rugosa fermentation for sustainable biodiesel. *Bioresource Technology Reports*, 2025. **29**.
30. Palanivel, P.D., *et al.*, Harnessing dark fermentation: Mechanisms, metabolic pathways, and nanoparticle innovations for biohydrogen enhancement. *International Journal of Hydrogen Energy*, 2025. **109**: p. 383–396.

31. Meisen, P. and T. Hammons. Harnessing the untapped energy potential of the oceans: Tidal, wave, currents and OTEC, in *2005 IEEE Power Engineering Society General Meeting*. 2005.

32. van Dijk, M., D. Gezer, and P. Rudolf. Unlocking hydropower's potential: Retrofitting infrastructure and harnessing unconventional sources for clean energy Transitions, in *IOP Conference Series: Earth and Environmental Science*. 2025. Institute of Physics.

33. Garrett, K.P., R.A. McManamay, and A. Witt, Harnessing the power of environmental flows: Sustaining river ecosystem integrity while increasing energy potential at hydropower dams. *Renewable and Sustainable Energy Reviews,* 2023. **173**.

34. Kiziroglou, M.E. and E.M. Yeatman, 17 - Materials and techniques for energy harvesting, in *Functional Materials for Sustainable Energy Applications,* J.A. Kilner, et al., Editors. 2012, Woodhead Publishing. p. 541–572.

35. Chen, Z., *et al.*, Recent progress of energy harvesting and conversion coupled with atmospheric water gathering. *Energy Conversion and Management,* 2021. **246**.

36. Sharghi, H. and O. Bilgen, Energy harvesting from human walking motion using pendulum-based electromagnetic generators. *Journal of Sound and Vibration*, 2022. **534**.

37. d'Angelo, M., C. Galassi, and N. Lecis, Thermoelectric materials and applications: A review. *Energies*, 2023. **16**(17).

38. Ashraf, M., *et al.*, Recent trends in sustainable solar energy conversion technologies: Mechanisms, prospects, and challenges. *Energy and Fuels*, 2023. **37**(9): p. 6283–6301.

39. Angenent, L.T., *et al.*, Electrical-energy storage into chemical-energy carriers by combining or integrating electrochemistry and biology. *Energy and Environmental Science*, 2024. **17**(11): p. 3682–3699.

40. Covaci, C. and A. Gontean, Piezoelectric energy harvesting solutions: A review. *Sensors (Switzerland)*, 2020. **20**(12): p. 1–37.

41. Carneiro, P., *et al.*, Electromagnetic energy harvesting using magnetic levitation architectures: A review. *Applied Energy*, 2020. **260**.

42. Moriarty, T.F., *et al.*, 4.407 - Bacterial Adhesion and Biomaterial Surfaces, in *Comprehensive Biomaterials*, P. Ducheyne, Editor. 2011, Elsevier: Oxford. p. 75–100.

43. Khan, F.U. and M.U. Qadir, State-of-the-art in vibration-based electrostatic energy harvesting. *Journal of Micromechanics and Microengineering*, 2016. **26**(10).

44. Elahi, H., M. Eugeni, and P. Gaudenzi, Chapter 3 - Energy harvesting, in *Piezoelectric Aeroelastic Energy Harvesting*, H. Elahi, M. Eugeni, and P. Gaudenzi, Editors. 2022, Elsevier. p. 41–59.

45. Feng, M., *et al.*, An overview of environmental energy harvesting by thermoelectric generators. *Renewable and Sustainable Energy Reviews*, 2023. **187**: p. 113723.

46. Collins, L., Harvest for the world. *IET Power Engineer*, 2006. **20**(1): p. 34–37.

47. Xie, W., *et al.*, Interactions of Melanin with Electromagnetic Radiation: From Fundamentals to Applications. *Chemical Reviews*, 2024. **124**(11): p. 7165–7213.

48. Huang, Y., *et al.*, A mechanically tunable electromagnetic wave harvester and dual-modal detector based on quasi-static van der Waals heterojunction. *Nano Energy*, 2022. **99**: p. 107399.
49. Ibrahim, H.H., *et al.*, Radio frequency energy harvesting technologies: A comprehensive review on designing, methodologies, and potential applications. *Sensors*, 2022. **22**(11).
50. Roy, S., *et al.*, Design of a highly efficient wideband multi-frequency ambient RF energy harvester. *Sensors*, 2022. **22**(2).
51. Munir, A., *et al.*, Metal nanoclusters: New paradigm in catalysis for water splitting, solar and chemical energy conversion. *ChemSusChem*, 2019. **12**(8): p. 1517–1548.
52. Hsu, W.S., *et al.*, Miniaturized salinity gradient energy harvesting devices. *Molecules*, 2021. **26**(18).
53. Zhou, X., *et al.*, Principles and materials of mixing entropy battery and capacitor for future harvesting salinity gradient energy. *ACS Applied Energy Materials*, 2022. **5**(4): p. 3979–4001.
54. Gul, H., *et al.*, Progress in microbial fuel cell technology for wastewater treatment and energy harvesting. *Chemosphere*, 2021. **281**.
55. Anastasovski, A., Is renewable energy storage sustainable? A review. *Green Technologies and Sustainability*, 2025. **3**(3).
56. Saha, M., *et al.*, A comprehensive review of novel emerging electrolytes for supercapacitors: Aqueous and organic electrolytes versus ionic liquid-based electrolytes. *Energy and Fuels*, 2024. **38**(10): p. 8528–8552.
57. Riaz, A., *et al.*, Review on comparison of different energy storage technologies used in micro-energy harvesting, wsns, low-cost microelectronic devices: Challenges and recommendations. *Sensors*, 2021. **21**(15).
58. Islam, M.M., *et al.*, Improving reliability and stability of the power systems: A comprehensive review on the role of energy storage systems to enhance flexibility. *IEEE Access*, 2024.**12**: 152738–152765.
59. Mensah-Darkwa, K., *et al.*, Supercapacitor energy storage device using bio-wastes: A sustainable approach to green energy. *Sustainability (Switzerland)*, 2019. **11**(2).
60. Wu, Z., *et al.*, Molecular and morphological engineering of organic electrode materials for electrochemical energy storage. *Electrochemical Energy Reviews*, 2022. **5**.
61. Phor, L., A. Kumar, and S. Chahal, Electrode materials for supercapacitors: A comprehensive review of advancements and performance. *Journal of Energy Storage*, 2024. **84**.
62. Shinde, P.A., *et al.*, Strengths, weaknesses, opportunities, and threats (SWOT) analysis of supercapacitors: A review. *Journal of Energy Chemistry*, 2023. **79**: p. 611–638.
63. Gao, M., *et al.*, Advances and challenges of green materials for electronics and energy storage applications: From design to end-of-life recovery. *Journal of Materials Chemistry A*, 2018. **6**(42): p. 20546–20563.
64. Liang, C., *et al.*, Novel semiconductor materials for advanced supercapacitors. *Journal of Materials Chemistry C*, 2023. **11**(13): p. 4288–4317.
65. Lipu, M.S.H., *et al.*, Battery management, key technologies, methods, issues, and future trends of electric vehicles: A pathway toward achieving sustainable development goals. Batteries, 2022. **8**(9).

66. Bajwa, R.A., *et al.*, Metal-organic framework (MOF) attached and their derived metal oxides (Co, Cu, Zn and Fe) as anode for lithium ion battery: A review. *Journal of Energy Storage*, 2023. **72**.

67. Patra, S., *et al.*, The positioning of biofuel cells-based biobatteries for net-zero energy future. *Journal of Energy Storage*, 2023. **72**.

68. Mu, T., *et al.*, Technological penetration and carbon-neutral evaluation of rechargeable battery systems for large-scale energy storage. *Journal of Energy Storage*, 2023. **69**.

69. Tafete, G.A., M.K. Abera, and G. Thothadri, Review on nanocellulose-based materials for supercapacitors applications. *Journal of Energy Storage*, 2022. **48**.

70. Takshi, A., *et al.*, A critical review on the voltage requirement in hybrid cells with solar energy harvesting and energy storage capability. *Batteries and Supercaps*, 2021. **4**(2): p. 252–267.

71. Atawi, I.E., *et al.*, Recent advances in hybrid energy storage system integrated renewable power generation: Configuration, control, applications, and future directions. *Batteries*, 2023. **9**(1).

72. Lin, Y., *et al.*, Controllable preparation of green biochar based high-performance supercapacitors. *Ionics*, 2022. **28**(6): p. 2525–2561.

Epilogue
The role of GSE in energy transition

9.1 INTRODUCTION

The world is currently addressing the facts of a changing climate. A major reason for this is the increased CO_2 emissions [1]. Energy harnessing, production, and use are major contributors to this situation. A means to address this is to work to achieve a carbon-neutral energy system. The path to this goal involves energy transition. This transition will not happen immediately, but it is occurring now. This transition will transform the present energy system from a predominantly fossil fuel-based system to a carbon-free energy system (see Figure 9.1).

Following decades of research and scientific discoveries, it is abundantly clear that anthropogenic carbon emissions from the use of fossil fuel resources are causing a rapid average warming of the earth's atmosphere [2].

Figure 9.1 Energy transition.

Source: Shutterstock Asset id: 2390387367.

DOI: 10.1201/9781003407447-12

Currently, the increase in greenhouse gas emissions has already resulted in an increase of 1.1 °C or more compared to preindustrial levels [3]. Increasing energy demands require new sources of energy; sources that can come from green sustainable energy (GSE) [4].

It is important to know where we have been energetically and transition to where we should go.

9.2 THE ENERGY TRANSITION

To reach a goal of net-zero carbon [5], the current energy system needs to change. At present, about 80% of the global primary energy supply is still being met with fossil energy [6]. To achieve the temperature goal, energy economies must shift from an energy mix based on fossil fuels to one that produces very limited, if not zero, carbon emissions, based on renewable energy sources. This is the energy transition that needs to occur.

The escalating greenhouse gas (GHG) emissions levels are forcing a global shift from carbon-intensive energy systems to sustainable and low-carbon emission energy alternatives (with associated technologies). Recognizing the critical role of the energy transition, it is important to address the factors propelling the transition in order to accomplish the desired reduction in CO_2 [7].

The "energy transition" is a transformation of the global energy sector from fossil-based systems of energy production and consumption to renewable energy sources. Switching from nonrenewable energy sources like oil, natural gas, and coal to renewable energy is made possible by technological advancements and a societal push toward sustainability. Spurred by structural, permanent changes to energy supply, demand, and prices, the energy transition also aims to reduce energy-related greenhouse gas emissions through various forms of decarbonization.

9.3 CURRENT ENERGY STATUS

The current global energy system is fully geared toward the supply chain and technical capabilities that fossil fuels enable [6]. This means that there is a well-developed energy system that balances supply and demand via a very liquid market. (This includes oil, coal, and natural gas.) This system is centralized, which means that the consumer only uses the energy and does not produce energy itself.

The world is moving toward green energy, but this does not mean that we can completely disregard fossil fuels. Fossil fuels are the most widely used energy sources in the world. Fossil fuels do have a value that must be considered, which favors its use:

- Easier to store and transport (than GSE)—they can be transported quite easily. They are typically transported via international gas pipeline systems or with tankers.
- It is relatively inexpensive, since it has been used for centuries, and a well-developed infrastructure for fossil fuels is in place.
- If the circumstances are good, and the weather is mild, the surplus can be stored, which enhances the yearly supply.
- It is currently more reliable than GSE.

9.4 THE PARADIGM SHIFT

Given the inevitability of an energy transition, what changes are needed before GSE assumes greater market share? GSE will not work within the traditional low-complexity central generation and distribution system. Most novel energy systems, such as solar, wind, geothermal, hydropower, and biogas, are smaller-scale distributed systems. With these new sources, consumers can also be producers, playing an increasingly important role when the sun is not shining or the wind is blowing too hard. This situation also demands flexible and large-scale storage to keep the supplies needed.

The shift away from fossil fuel systems to renewable and low-carbon systems will require the distribution and storage network to transition as well to a complex multiple generation and distribution system. The resulting energy system will become a much more flexible system with a higher degree of intelligence and responsiveness [8]. Even in this system fossil fuels will continue to play an important role. The use of these fuels will be necessary to provide a bridge until sufficient and acceptable GSE are developed.

9.5 NEW PATH FORWARD

Current energy systems supply a variety of forms and power. In most systems around the world, electricity is only a small component, and fossil liquid and gaseous fuels make up the largest part of the total energy system [6]. Included in these are specialized needs (like jet fuels) that are being developed [9]. Current technologies to decarbonize the energy system favor electricity generation, and therefore the trend is toward further electrification of sectors like transport, households, and industry.

The more the electricity can meet energy needs, the faster the energy transition will occur. The complete electrification of the demand side will only require decarbonization of the electricity generation supply side. Switching from fossil fuel sources toward solar and wind assets can be done more easily, especially since the costs of these renewable energy systems have come down significantly. They are currently on par with or below fossil fuel-generated electricity [10]. Until sufficient GSE is developed, the alternative is the use of carbon capture and storage (CCS) to capture or store the CO_2 produced from the use of fossil fuels.

The energy transition is a process of moving away from fossil fuels as we know them. That does mean a gradual transition away from fossil fuels to low-emission fossil fuels and GSE. This process will take decades, providing a role for scientists, economists, and politicians during the transition period as well as the post-transition time. Additionally, the move to a GSE-based economy will require significant development to allow such an economy to function.

REFERENCES

1. Stille, L., M. Wilson, and K. Bishop, *Introduction to the energy transition*, in *Geophysics and the Energy Transition*. 2024, Elsevier. 3–13.
2. Jagger, P., *et al.*, SDG 7: Affordable and clean energy-How access to affordable and clean energy affects forests and forest-based livelihoods, in *Sustainable Development Goals: Their Impacts on Forests and People*. 2019, Cambridge University Press. p. 206–236.
3. Syvitski, J., *et al.*, Extraordinary human energy consumption and resultant geological impacts beginning around 1950 CE initiated the proposed Anthropocene Epoch. *Communications Earth and Environment*, 2020. **1**(1).
4. Mahmud, M.A.P., *et al.*, Green energy: A sustainable future, in *Green Energy: A Sustainable Future*. 2023: Elsevier. 1–239.
5. Sharma, V.K., *et al.*, A comprehensive review of green energy technologies: Towards sustainable clean energy transition and global net-zero carbon emissions. Processes, 2025. **13**(1).
6. IEA, *World Energy Outlook 2024*, P. IEA, Editor. 2024.
7. Muhire, F., *et al.*, Drivers of green energy transition: A review. *Green Energy and Resources*, 2024. **2**(4).
8. Oduro, R.A. and P.G. Taylor, Future pathways for energy networks: A review of international experiences in high income countries. *Renewable and Sustainable Energy Reviews*, 2023. **171**.
9. Ozkan, M., *et al.*, Forging a sustainable sky: Unveiling the pillars of aviation e-fuel production for carbon emission circularity. *iScience*, 2024. **27**(3).
10. Agarwala, A., *et al.*, Towards next generation power grid transformer for renewables: Technology review. *Engineering Reports*, 2024. **6**(4).

Index

For Product Safety Concerns and Information please contact our EU
representative GPSR@taylorandfrancis.com
Taylor & Francis Verlag GmbH, Kaufingerstraße 24, 80331 München, Germany

www.ingramcontent.com/pod-product-compliance
Lightning Source LLC
Chambersburg PA
CBHW052013230326
41598CB00078B/3218